T0220391

Lecture Notes in Mathematics

An informal series of special lectures, seminars and reports on mathematical topics

Edited by A. Dold, Heidelberg and B. Eckmann, Zürich

8

Gaetano Fichera

University of Rome

Linear elliptic differential systems and eigenvalue problems

The Johns Hopkins University, Baltimore Md, March – May 1965

1965

Springer-Verlag · Berlin · Heidelberg · New York

Library of Congress Catalog Card Number 65—27796. Printed in Germany. Title No. 7328

These Notes contain the lectures I had the pleasure of delivering as Visiting Professor at the Department of Mechanics of The Johns Hopkins University on the invitation of Professor Clifford Truesdell.

They are intended to be an introduction to the modern approach to higher order elliptic boundary value problems and related eigenvalue problems.

I am deeply grateful to Dr. Warren Edelstein for his kind collaboration in checking both the English and the Mathematics of these Notes.

G. Fichera

Baltimore, Md. - May 1965.

C O N T E N T S

Lecture 1

"Well posed" boundary value problems.

The classical point of view in the theory of partial differential equations (PDE) assumes that such equations admit to an infinite number of solutions. The typical problem connected with a PDE consists of finding amongst all those possible solutions a particular one which satisfies properly given auxiliary conditions. These conditions are generally expressed as complementary equations which the unknown functions must satisfy on the boundary of the domain, where the PDE is considered, or on part of it. For this reason, these conditions are known as "boundary conditions".

In spite of the fact that this point of view is the one that agrees with applications of PDE in various branches of applied mathematics, it must be pointed out that the assumption that a PDE, even a linear one with very smooth coefficients, possesses infinitely many solutions may be false, since the equation might fail to have any solution. In this respect let us consider a very interesting example given by Hans Lewy [4] a few years ago.

Let us consider, in the real cartesian 3-space with coordinates x, y, t, the PDE

$$Lu \equiv \frac{1}{2}\left(\frac{\partial u}{\partial x} + i\frac{\partial u}{\partial y}\right) + iz\frac{\partial u}{\partial t} = f(x,y,t)$$

where $z = x + iy$. By using the Wirtinger differential operator

$\frac{\partial u}{\partial \bar{z}} = \frac{1}{2}\left(\frac{\partial u}{\partial x} + i\frac{\partial u}{\partial y}\right)$, we can write Lu in a more compact form

$$Lu \equiv \left(\frac{\partial}{\partial \bar{z}} + iz\frac{\partial}{\partial t}\right)u.$$

We assume that $f(x,y,t)$ is a function depending only on t, which we suppose to be real valued. We write, for convenience, such a function as a derivative of a real function $\psi(t)$. The above equation can then be written

$$(1.1) \qquad \left(\frac{\partial}{\partial \bar{z}} + iz\frac{\partial}{\partial t}\right)u = \frac{d\psi}{dt}.$$

We shall prove that <u>a necessary condition for (1.1) to have a continuously differentiable solution in a neighborhood of the origin is that ψ be a real analytic function of t</u>.

Let r be a positive real number and set $z = r^{\frac{1}{2}}e^{i\theta}$. It is easily seen that

$$\frac{\partial u}{\partial \bar{z}} = z\frac{\partial u}{\partial r} + \frac{i}{2}\frac{z}{r}\frac{\partial u}{\partial \theta}.$$

Since

$$\frac{i}{2}\int_0^{2\pi}\frac{z}{r}\frac{\partial u}{\partial \theta}\,d\theta = \int_0^{2\pi}u\frac{\partial z}{\partial r}\,d\theta,$$

we obtain

$$(1.2) \qquad \int_0^{2\pi}\frac{\partial u}{\partial \bar{z}}\,d\theta = \frac{\partial}{\partial r}\int_0^{2\pi}zu\,d\theta.$$

From (1.1) and (1.2), by assuming $z = r^{\frac{1}{2}}e^{i\theta}$ and integrating we get

$$(1.3) \qquad \left(\frac{\partial}{\partial r} + i\frac{\partial}{\partial t}\right)\int_0^{2\pi}r^{\frac{1}{2}}e^{i\theta}u(r^{\frac{1}{2}}e^{i\theta},t)\,d\theta = 2\pi\frac{d\psi}{dt}.$$

Now set $\zeta = r + it$ and

$$U(\zeta) = \int_0^{2\pi} r^{\frac{1}{2}} e^{i\theta} u(r^{\frac{1}{2}} e^{i\theta}, t)\, d\theta .$$

Equation (1.3) gives

$$\left(\frac{\partial}{\partial r} + i \frac{\partial}{\partial t}\right)\left(U + i 2\pi \psi\right) = 0.$$

This means that the function $V = U + i2\pi\psi$ is an holomorphic function of ζ for $0 < r < \varepsilon$ and $-\varepsilon < t < \varepsilon$ with ε conveniently chosen. The function $U(\zeta)$ is continuous for $0 \leq r < \varepsilon$ and vanishes for $r = 0$. From this it follows that the real part of V is continuous for $0 \leq r < \varepsilon$ and vanishes for $r = 0$. Therefore V can be continued analitically across the t-axis of the (r,t)-plane. Hence ψ - as the trace of V on the t-axis - is analytic.

If the real function ψ is as smooth as we want, <u>but not analytic</u>, equation (1.1) has no solution in any arbitrarily fixed neighborhood of the origin.

Let us now consider another example of a linear PDE of the first order which possesses only one solution in a given domain. Let Q be the closed square of the (x,y)-plane

$$-1 \leq x \leq 1 \qquad , \qquad -1 \leq y \leq 1$$

and let $b_1(x,y)$ and $b_2(x,y)$ be two real functions defined on Q, as smooth as we wish and satisfying the following conditions

$$b_1(-1,y) \geq 0 \qquad , \qquad b_1(1,y) \leq 0,$$

$$b_2(x,-1) \geq 0 \qquad , \qquad b_2(x,1) \leq 0.$$

Let $c(x,y)$ be an arbitrarily smooth real function defined on Q and negative at every point of Q . We want to prove that the only continuous differentiable solution in the square Q of the linear PDE

$$(1.4) \qquad b_1(x,y)\frac{\partial u}{\partial x} + b_2(x,y)\frac{\partial u}{\partial y} + c(x,y)u = 0$$

is the trivial solution $u \equiv 0$. In fact, the minimum m of the real solution u of (1.4) over Q cannot be negative. If m is taken on in an interior point of Q , then, obviously, $m = 0$.
Let us assume that u takes on its minimum m on the boundary, for instance in the point $(x_0,-1)$. Then $u_y(x_0,-1) \geq 0$ and therefore $b_2(x_0,-1)u_y(x_0,-1) \geq 0$. $u_x(x_0,-1) = 0$ for $-1 < x_0 < 1$ and $u_x(x_0,-1) \geq 0$, if $x_0 = -1$ and $u_x(x_0,-1) \leq 0$ if $x_0 = 1$. In any case $b_1(x_0,-1)u_x(x_0,-1) \geq 0$. Since it must be the case that $c(x_0,-1)u(x_0,-1) \leq 0$, it follows $m = u(x_0,-1) \geq 0$. The function $-u$ is a solution of (1.4) and therefore the maximum of u over Q cannot be positive. Then $u \equiv 0$.

This example, which was given by Mauro Picone [5] ,has been known since 1928. More sophisticated examples of homogeneous equations of higher order whose only solution is $u \equiv 0$ can also be constructed (see [2],[3]).

These simple examples show that, besides the classical point of view, one has to include also situations like the above considered ones, when aiming to describe what a "well posed boundary value problem" is for a general system of linear PDE.

Let A be a domain (i.e. connected open set) of the real cartesian space X^r. The point of X^r will be denoted by $x \equiv (x_1,\dots,x_r)$. We shall consider vector-valued functions $u(x)$ defined on A . More precisely, when we say that $u(x)$ is an n-vector function, we mean that the values of u are n-vectors of the n-dimensional complex cartesian space. By $D \equiv (D_1,\dots,D_r)$, $D_i = \frac{\partial}{\partial x_i}$ we denote the differentiation vector" The letters $p, q, s, \ldots$ will denote r-vectors with non-negative integral components e.g., $p \equiv (p_1, \ldots, p_r)$, and we set $|p| = \sum_{i=1}^{r} p_i$. Otherwise for any vector $u \equiv (u_1, \ldots, u_m)$, $|u|$ will represent its Cartesian lenght $|u|^2 = \sum_{i=1}^{m} |u_i|^2$ and $uv = \sum_{i=1}^{m} u_i \bar{v}_i$. We write $[\xi \equiv (\xi_1, \ldots, \xi_r)]$:

$$\xi^p = \xi_1^{p_1} \ldots \xi_r^{p_r}, \quad D^p = D_1^{p_1} \ldots D_r^{p_r} \qquad (\xi_i^{p_i} = 1 \text{ if } \xi_i = p_i = 0).$$

If B is any point-set of X^r with interior points, we shall denote by $C^k(B)$ the class of all the (vector-valued-) functions u possessing continuous derivatives up to the order k in B. This means that any derivative of u of order $\leq k$ exists at every interior point of B and coincides with a function which is continuous in the whole set B.
If u is any function defined on A, we denote by spt u (support of u) the closure of the set where $|u(x)| > 0$. $\mathring{C}^k(A)$ will denote the sub-class of $C^k(A)$ consisting of functions u such that spt $u \subset A$.
By C^k we shall denote the class of functions defined in the whole space X^r and possessing continuous derivatives up to the order k, i. e. $C^k = C^k(X^r)$. $\mathring{C}^k$ will denote the sub-class of C^k consisting of functions with a bounded support. The symbols $C^\infty(A)$, $\mathring{C}^\infty(A)$, C^∞, $\mathring{C}^\infty$ are self explanatory.
Let $a_s(x)$ be an $m \times n$-matrix defined on A $(0 \leq |s| \leq \nu)$. We denote by Lu the matrix differential operator

$$Lu \equiv a_s D^s u.$$

We use here the summation convention, i.e. when a vector-index s is repeated twice, a summation must be understood which is extended to the whole domain of variability of s.
Let $\{S_h\}$ be a discrete set of complex vector spaces. By discrete we mean the set to be empty, finite, or countable. Let M_h be a linear

transformation defined on $C^{\nu}(\bar{A})$ and with range in the vector space S_h.

We shall consider the following problem

(1.5) $Lu = f$, (1.6) $M_h u = g_h$ $(h = 1, \dots)$.

The symbol f denotes a given m-vector valued function defined on A, g_h a given vector of the space S_h.
In spite of their extremely abstract definition we shall refer to conditions (1.6) (when $\{S_h\}$ is not empty) as boundary conditions.
In the case that $\{S_h\}$ is empty, the problem consists merely in finding a solution u of the equation (1.5).

Let us furthermore suppose that A is a bounded regular domain of X^r (i.e. the Green-Gauss identity holds for it) and that the matrix $a_s(x)$ belongs to $C^{|s|}(A)$. If $a_s \equiv ((a_s^{hk}))$ $(h = 1, \dots, m; k = 1, \dots, n)$ we denote by $\bar{a}_s$ the adjoint matrix, i.e. the $n \times m$ matrix $((\bar{a}_s^{kh}))$. The following matrix-differential operator will be called the adjoint operator of L

$$L^* v = (-1)^{|s|} D^s \bar{a}_s v .$$

Suppose that u and v both belong to $C^{\nu}(\bar{A})$ then the following Green's formula holds:

$$\int_A u L^* v \, dx - \int_A f v \, dx = \int_{\partial A} H(u, v) \, d\sigma$$

where $H(u,v)$ is a bilinear [1] differential operator of order $\nu - 1$

[1] Since $H(u,v)$ is defined for complex vector valued functions the term "bilinear" means that $H(u,v)$ is linear with respect to u, i.e. $H(au + bu', v) = aH(u,v) + bH(u',v)$ and antilinear with respect to v, i.e. $H(u, av + bv') = \bar{a} H(u,v) + \bar{b} H(u,v')$ $[\bar{a}, \bar{b}$ are the complex conjugates of a and $b]$.

in u and in v, whose coefficients are expressed in terms of the matrices a_s and of the first order differential elements of the boundary ∂A. It is somewhat tedious to write down explicitly the full expression of $H(u,v)$. However, this is not needed for our purposes.

We shall consider, instead of the general problem (1.5),(1.6) the following one with "homogeneous boundary conditions"

(1.5) $\quad Lu = f \quad , \qquad$ (1.6$_0$) $\quad M_h u = 0.$

When a function u^0 (belonging to the space of what we shall define as admissible solutions) exists such that $M_h u^0 = g_h$, then, and only then, problem (1.5),(1.6) is equivalent to problem (1.5),(1.6$_0$).
Let us denote by V the linear variety consisting of all m-vector valued functions belonging to $C^\nu(\bar{A})$ such that

$$\int_{\partial A} H(u,v)\, d\sigma = 0$$

for any u satisfying conditions (1.6$_0$) and belonging to $C^\nu(\bar{A})$.
If a solution u of problem (1.5), (1.6$_0$) exists belonging to $C^\nu(\bar{A})$ then u must satisfy the integral equation

$$(1.7) \qquad \int_A u\, L^* v\, dx = \int_A f v\, dx$$

for every $v \in V$. This is the starting point of the concept of <u>weak solution</u> for the boundary problem (1.5),(1.6$_0$). It consists merely in substituting the integral equations (1.7), written for any $v \in V$, for the equations (1.5),(1.6$_0$). In order to make the equations (1.7) consistent we assume the following hypothesis.

1°) <u>The linear variety</u> V <u>contains some vector different from the zero-vector.</u>

It is convenient to enlarge our problem in order to include the possibility that the given function f and the unknown function u be generalised functions. We do this in a quite abstract way.

Let S_u be a complex Banach space ($\mathcal{B}$-space). We assume that S_u contains a linear subvariety that is linear-isomorphic to $C^{\nu}(\bar{A})$. S_u will be the space of the admissible unknowns. Let S_f be a second $\mathcal{B}$-space. We assume that S_f contains a linear subvariety linear-isomorphic to $C^{0}(\bar{A})$.

In addition to the hypothesis 1°) we make the following ones:

2°) There exist two complex $\mathcal{B}$-spaces $\mathcal{B}_u$ and $\mathcal{B}_f$ such that $\mathcal{B}_u$ consists of measurable (complex n-vector valued) functions and $\mathcal{B}_f$ of measurable (complex m-vector valued) functions. Moreover $S_u = \mathcal{B}_u^*$ and $S_f = \mathcal{B}_f^*$ ($\mathcal{B}_u^*$ and $\mathcal{B}_f^*$ are the topological dual spaces of $\mathcal{B}_u$ and $\mathcal{B}_f$ respectively).

3°) $\mathcal{B}_f$ contains V and $\mathcal{B}_u$ contains the range of L^*v when v varies in V.

4°) If S_u $[S_f]$ contains a linear subvariety Banach-isomorphic to a $\mathcal{B}$-space of measurable functions, then, if u $[f]$ denotes any function of this subvariety and w is any function of $\mathcal{B}_u$ $[\mathcal{B}_f]$, the scalar function uw $[fw]$ is Lebesgue integrable on A and

$$\langle u, w\rangle = \int_A u\,w\,dx \qquad \left[\langle f, w\rangle = \int_A f\,w\,dx\right].$$

($\langle\ \rangle$ denotes the duality between a $\mathcal{B}$-space and its topological dual space).

We shall consider the following problem:

A vector f of the $\mathcal{B}$-space S_f is given. We want to find a vector u of the $\mathcal{B}$-space S_u such that

$$\langle u, L^* v\rangle = \langle f, v\rangle \tag{1.8}$$

for any $v \in V$.

Because of hypothesis 4°), when f and u are functions in the classical sense, then the system (1.8) reduces to the system (1.7).

The vector u will be called a <u>weak solution</u> of the boundary value problem (1.5),(1.6$_0$), with space S_f as the space of "<u>data</u>" and space S_u as the space of <u>admissible solutions</u>.

Assume that there exists some nontrivial solution of the equation $L^* v$ in the variety V. Let V° be the linear subvariety of V consisting of all these solutions, then a necessary condition for the existence of a solution of our problem is that

$$(1.9) \qquad \langle f, v^\circ \rangle = 0 \qquad \text{for} \qquad v^\circ \in V^\circ.$$

We shall say that the boundary value problem (B.V.P.) (1.5),(1.6$_0$) is a <u>well posed boundary value problem in the spaces</u> S_f, S_u, when for any $f \in S_f$ satisfying the compatibility condition (1.9), there exists some vector $u \in S_u$ satisfying equations (1.8).

We want now to give a necessary and sufficient condition for a B.V.P. to be well posed. Let $\bar{V}^\circ$ be the closure of the variety V° in the B-space B_f. Let us denote by $\mathcal{F}$ the factor B-space $B_f / \bar{V}^\circ$. If w is any function in B_f, we shall denote by $[w]$ the equivalence class - as an element of $\mathcal{F}$ - determined by w. Set

$$\Delta = \inf_{v \in V - \bar{V}^\circ} \frac{\| L^*(v) \|_{B_u}}{\| [v] \|_{\mathcal{F}}}.$$

In the next lecture we shall prove the following theorem:

1.I. <u>A necessary and sufficient condition for the B.V.P. (1.5), (1.6$_0$) to be well posed in the spaces</u> S_f, S_u <u>is that</u> Δ <u>be greater than zero</u>.

Δ will be called the <u>discriminator</u> of the B.V.P. (1.5),(1.6$_0$) in the spaces S_f, S_u.

Bibliography of Lecture 1

[1] G. FICHERA - Lezioni sulle trasformazioni lineari - Ediz. Veschi - Trieste, 1954.

[2] G. FICHERA - Premesse ad una teoria generale dei problemi al contorno per le equazioni differenziali - Corsi INAM Ediz. Veschi - Roma,1958.

[3] G. FICHERA - Sul concetto di problema "ben posto" per una equazione differenziale - Rendiconti di Matematica - 19 - 1960.

[4] H. LEWY - An example of a smooth linear partial differential equation without solution - Annals of Mathematics - 66 - 1957.

[5] M. PICONE - Maggiorazione degli integrali delle equazioni totalmente paraboliche alle derivate parziali del secondo ordine - Annali di Matematica pura e applicata - 1929.

Lecture 2

Existence principle.

Let V be a complex vector space and B_1 and B_2 two complex B-spaces. Let M_h $(h = 1, 2)$ be a linear transformation with domain V and range in the space B_h. We denote by B_h^* the topological dual space of B_h.

We shall consider the following problem:

A vector φ of the space B_1^* is given; find a vector ψ of B_2^* such that

(2.1) $\langle \varphi, M_1 v \rangle = \langle \psi, M_2 v \rangle$ for any $v \in V$.

We shall prove the following theorem:

2.I. A necessary and sufficient condition for the existence of a solution of problem (2.1), for any $\varphi \in B_1^*$, is that a positive constant κ exist such that the following inequality hold for any $v \in V$:

(2.2) $$\| M_1 v \| \leq \kappa \| M_2 v \| .$$

Sufficiency. Let w_2 be any vector in the range $M_2(V)$ of M_2. Let $w_2 = M_2 v$ and set $w_1 = M_1 v$. The vector w_1 is uniquely determined by w_2, since $M_2 v = M_2 v' = w_2$ implies, because of (2.2), that $\| M_1 v - M_1 v' \| \leq \kappa \| M_2 v - M_2 v' \| = 0$. Let us define on $M_2(V)$ the functional

$$\psi(w_2) = \langle \varphi, w_1 \rangle \equiv \langle \varphi, M_1 v \rangle .$$

Obviously Ψ depends linearly on w_2. On the other hand, $|\Psi(w_2)| \leq \|\varphi\| \|M_1 v\| \leq K \|\varphi\| \|M_2 v\| = K \|\varphi\| \|w_2\|$. Therefore Ψ is a bounded functional of w_2 and

$$(2.3) \qquad \|\Psi\| \leq K \|\varphi\|.$$

By the Hahn-Banach theorem we may continue Ψ in the whole space B_2 in such a way that (2.3) still holds for the continued functional. Such a functional will be a solution of (2.2).

Necessity. We can restrict ourselves to consideration of the subspaces $\overline{M_1(V)}$ and $\overline{M_2(V)}$ of B_1 and B_2 respectively, instead of the whole spaces B_1 and B_2. This allows us to say that for any given $\varphi \in [\overline{M_1(V)}]^*$ there exists only one $\Psi \in [\overline{M_2(V)}]^*$ that is a solution of (2.1). Let us denote by $\Psi = T\varphi$ the solution of (2.1) corresponding to φ. T is a linear transformation of $[\overline{M_1(V)}]^*$ into $[\overline{M_2(V)}]^*$. T has a closed graph. In fact, let us suppose that $\varphi_n \to \varphi$ and $T\varphi_n \to \Psi$. Then we have $\langle \varphi_n, M_1 v\rangle \to \langle \varphi, M_1 v\rangle$ and $\langle T\varphi_n, M_2 v\rangle \to \langle \Psi, M_2 v\rangle$. Since $\langle \varphi_n, M_1 v\rangle = \langle T\varphi_n, M_2 v\rangle$, then $\langle \varphi, M_1 v\rangle = \langle \Psi, M_2 v\rangle$, i.e. $\Psi = T\varphi$. By the "closed graph theorem", T is a bounded transformation. A constant K exists such that $\|\Psi\| = \|T\varphi\| \leq K \|\varphi\|$. For any fixed w_2 in the range $M_2(V)$, let us consider the functional $W_{w_2}(\varphi) = \langle T\varphi, w_2\rangle$. This functional is linear and it is bounded since $|W_{w_2}(\varphi)| = |\langle \Psi, w_2\rangle| \leq K \|\varphi\| \|w_2\|$. Henceforth $\|W_{w_2}\| \leq K \|w_2\|$. Let us now consider for any w_1 in the range $M_1(V)$, the linear functional $W_{w_1}(\varphi) = \langle \varphi, w_1\rangle$. It is bounded and $\|W_{w_1}\| \leq \|w_1\|$. Also let us consider the linear subvariety U of $M_1(V)$ consisting of any u such that $u = t w_1$ (t is a complex constant). Let us define on U the functional $f(u) = t \|w_1\|^2$. f is linear. It is bounded since

$$\frac{|f(u)|}{\|u\|} = \frac{|t|\,\|w_1\|^2}{|t|\,\|w_1\|} = \|w_1\|$$

and $\|f\| = \|w_1\|$. By the Hahn-Banach theorem, there exists a linear bounded functional F defined on $\overline{M_1(V)}$ such that $\|F\| = \|w_1\|$, $\langle F, u \rangle = f(u)$ for $u \in U$. We have $|W_{w_1}(F)| = |\langle F, w_1 \rangle| = \|w_1\|^2$ and also: $|\langle W_{w_1}, F \rangle| \leq \|W_{w_1}\|\,\|F\| = \|W_{w_1}\|\,\|w_1\|$. Therefore $\|W_{w_1}\| \geq \|w_1\|$. We have proved so far that $\|W_{w_1}\| = \|w_1\|$. Let us now assume $w_1 = M_1 v$, $w_2 = M_2 v$. Then we have, by (2.1), $W_{w_2}(\varphi) = W_{w_1}(\varphi)$ for any $\varphi \in [\overline{M_1(V)}]^*$. It follows that $\|W_{w_1}\| = \|W_{w_2}\|$. Since $\|W_{w_1}\| = \|w_1\| = \|M_1 v\|$ and $\|W_{w_2}\| \leq K\,\|w_2\| = K\,\|M_2 v\|$, (2.2) follows.

<u>Remark</u>. From the proof of the first part of the theorem, it follows that a solution exists satisfying the inequality (2.3) with the same constant K that appears in (2.2). The space of the <u>eigensolutions</u> of (2.1)is obviously composed of the functionals of B_2^* that are orthogonal to the range of M_2. This space is a closed subspace of B_2^*. Let us denote it by Ψ° and consider the factor space B_2^* / Ψ°. We shall denote by $[\psi]$ the equivalence class – as an element of this factor space – determined by ψ. Since the norm $\|[\psi]\|$ in the factor space is defined as follows

$$\|[\psi]\| = \inf_{\psi^\circ \in \Psi^\circ} \|\psi + \psi^\circ\|$$

as a consequence of (2.3) we have that, for any solution of (2.1), the following inequality holds

(2.4) $$\| [\psi] \| \leq K \| \varphi \| .$$

Inequality (2.4) will be called the <u>dual inequality</u> of (2.2).

It is evident that in order for the existence principle just proven to be usable it is necessary that the kernel V_2 of the transformation M_2 be contained in the kernel V_1 of M_1, i.e. $M_2 v = 0$ must imply that $M_1 v = 0$. In the opposite case, i.e. $V_2 \not\subset V_1$, the given vector φ cannot be chosen arbitrarily and the following necessary condition for the existence of ψ must be satisfied

(2.5) $$\langle \varphi, M_1 v_2 \rangle = 0 \qquad \text{for every} \quad v_2 \in V_2 .$$

We wish now to generalize theorem 2.I. and to give the necessary and sufficient condition for the existence of a solution of equation (2.1) for any φ satisfying the condition (2.5).

Let us consider the closure of the linear variety $M_1(V_2)$ and the Banach factor space $Q = B_1 / \overline{M_1 V_2}$. As usual, the norm of an equivalence class $[w]$ of Q is given by

$$\| [w] \| = \inf_{v_2 \in V_2} \| w + M_1 v_2 \| .$$

Let us denote by $\tilde{M}_1$ the linear transformation with domain V and range in the space Q that brings v into the equivalence class $[M_1 v]$ of the space B_1. The following theorem holds:

2.II. <u>A necessary and sufficient condition for the existence of a solution ψ of (2.1), for any φ satisfying (2.5), is that a positive constant K exist such that</u>

(2.6) $$\| \tilde{M}_1 v \| \leq K \| M_2 v \| .$$

This theorem, which appears as a generalisation of theorem 2.I.,can be considered as a particular case of that theorem (1). In fact; it is known that, if Φ is an element of Q^*, then it admits the following representation:

(2.7) $$\langle \Phi , [u] \rangle = \langle \varphi, u \rangle$$

where φ is an element of B_1^* vanishing identically on $M_1(V_2)$.Conversely, if φ is such a functional, then (2.7) defines an element of Q^*.Because of the condition (2.5), we may write (2.1) as follows

$$\langle \Phi , \tilde{M}_1 v \rangle = \langle \psi, M_2 v \rangle ,$$

where Φ is defined by (2.7). Since $M_2 v = 0$ implies $\tilde{M}_1 v = 0$, we can apply theorem 2.I. Thus, we get the proof.

Since $\|\Phi\| = \|\varphi\|$, the dual inequality of (2.6) is still (2.4).

It is evident how theorem 1.I is a particular case of theorem 2.II. In fact, let us assume that the vector space V is the linear variety $\bar{V}$ introduced in lecture 1 and that $B_1 = B_f$ and $B_2 = B_u$. Then the condition $\Delta > 0$ simply expresses the fact that (2.6) holds.

If the B.V.P. (1.5),(1.6$_0$) is well posed in the spaces S_f , S_u , then there exists a constant K such that the dual inequality holds:

(2.8) $$\| [u] \| \leq K \| f \| .$$

$[u]$ is the equivalence class of (weak) solutions of the B.V.P., when considered in the spaces S_f , S_u .
Inequality (2.8) expresses the fact that, when a B.V.P. is well posed, the equivalence class of the solutions depends continuously on the given data of the problem.

The classic Hadamard point of view in defining a well posed B.V.P.

assumes existence and continuous dependence of the solution on the given data . From our approach it follows that the second condition is a consequence of the first.

Bibliography of Lecture 2.

[1] N.DUNDFORD-J.SCHWARTZ - Linear operators - Interscience Publ. Inc. New York-London, 1958.

[2] G.FICHERA - see [1] , [2] , [3] , in Lecture 1.

[3] G.FICHERA - Operatori di Riesz-Fredholm, operatori riducibili, etc - Istituto Matematico Guido Castelnuovo - Roma,1964.

[4] E.HILLE-R.PHILLIPS - Functional Analysis and Semi-Groups - Amer.Math. Soc. Coll. Publ. 31 - Providence,1957.

[5] F.RIESZ-B.Sz NAGY - Leçons d'Analyse fonctionelle - Akademiai Kiado, Budapest, 1953 - English Translation, Ungar, New York,1955.

[6] A.C.ZAANEN - Linear Analysis - North Holland Publ. Co. Amsterdam,1953.

Lecture 3

The function spaces $\overset{\circ}{H}_m$ and H_m.

Let A be a bounded domain of the real cartesian space X^n. Without any loss in generality, we may assume that $\bar{A}$ is contained in the open square $Q : |x_k| < \pi$. Otherwise, we may use a proper change of coordinates in order to reduce to this case. Let us consider the vector space $\overset{\circ}{C}{}^m(A)$ $[C^m(\bar{A})]$ of complex n-vector valued functions, and for any pair u, v of vectors of this space let us define the following scalar product

$$(3.1) \qquad (u, v)_m = \int_A D^s u \, \overline{D^s v} \, dx. \qquad (0 \leq |s| \leq m).$$

We define spaces $\overset{\circ}{H}_m(A)$ $\{H_m(A)\}$ or simply (when any confusion can be avoided) $\overset{\circ}{H}_m$ $\{H_m\}$ to be the Hilbert function spaces obtained by functional completion of $\overset{\circ}{C}{}^m(A)$ $[C^m(\bar{A})]$ with respect to the scalar product (3.1)

It is obvious that for $m = 0$ we have $\overset{\circ}{H}(A) = H_0(A) = \mathcal{L}^2(A)$. In general, a function u will belong to $\overset{\circ}{H}_m$ (H_m) if and only if there exists a sequence of vectors of $\overset{\circ}{C}{}^m(A)$ $\{C^m(\bar{A})\}$, say $\{v_k(x)\}$, such that $\lim_{k \to \infty} \int_A |v_k(x) - u(x)|^2 dx = 0$ and, moreover, there exist functions $\varphi^s(x)$ $(0 < |s| \leq m)$ of $\mathcal{L}^2(A)$ such that $\lim_{k \to \infty} \int_D |D^s v_k(x) - \varphi^s(x)|^2 dx = 0.$ The function $\varphi^s(x)$ - which obviously does not depend on the sequence $\{v_k\}$ - is called the <u>strong derivative</u> D^s of the function u of $\overset{\circ}{H}_m$ (H_m), or, simply, the s-<u>derivative</u> of u.

It is easy to see that for any $u \in \overset{\circ}{H}_m$ and any $w \in H_m(A)$ the integration by parts formula holds:

(3.2) $$\int_A u D^s w\, dx = (-1)^{|s|} \int_A (D^s u) w\, dx \quad , (0 < |s| \leq m).$$

In fact (3.2) holds when we replace u by v_κ and w by z_κ ($\{v_\kappa\}$ and $\{z_\kappa\}$ are the above mentioned sequences with $v_\kappa \in \overset{\circ}{C}{}^m(A)$) and then, making κ tend to infinity, we obtain (3.2).

Any function u of $\overset{\circ}{H}_m(A)$ can be developed in a Fourier trigonometrical series

(3.3) $$u(x) = \frac{1}{(\sqrt{2\pi})^r} \sum_{-\infty}^{+\infty}{}_\kappa\, c_\kappa e^{i\kappa x}$$

where

$$c_\kappa = \frac{1}{(\sqrt{2\pi})^r} \int_A u e^{-i\kappa x} dx ,$$

the development being convergent in $\mathcal{L}^2(A)$. For $|s| \leq m$ we have also:

$$D^s u(x) = \frac{1}{(2\pi)^r} \sum_{-\infty}^{+\infty}{}_\kappa\, e^{i\kappa x} \int_A (D^s u) e^{-i\kappa x} dx$$

and then, using (3.2):

(3.4) $$D^s u(x) = \frac{1}{(\sqrt{2\pi})^r} \sum_{-\infty}^{+\infty}{}_\kappa\, (i)^{|s|} \kappa^s c_\kappa e^{i\kappa x}.$$

This means that for any $u \in \overset{\circ}{H}_m$, the Fourier series (3.3) can be differentiated term by term provided the order of the differentiation does not exceed m. We have:

$$\int_A |D^s u|^2 dx = \sum_{-\infty}^{+\infty}{}_\kappa\, (\kappa^s)^2 |c_\kappa|^2.$$

There exist two positive numbers p_0 and p_1 such that

$$p_0 |\kappa|^{2\nu} \leq \sum_{|s|=\nu} (\kappa^s)^2 \leq p_1 |\kappa|^{2\nu} \qquad (|\kappa| = \sqrt{\kappa_1^2 + \dots + \kappa_r^2}\,)$$

Therefore,

$$(3.5)\quad p_0 \sum_{-\infty}^{+\infty}{}_{\kappa} |\kappa|^{2\nu} |c_\kappa|^2 \le \sum_{|\delta|=\nu} \int_A |D^\delta u|^2 dx \le p_1 \sum_{-\infty}^{+\infty} |\kappa|^{2\nu} |c_\kappa|^2 .$$

Let us denote by $\|u\|_m$ the norm of an element of $\mathring{H}_m$ (or H_m). The following __lemma of the Poincarè inequality__ holds:

3.I. __For any__ $u \in \mathring{H}_m(A)$,

$$(3.6)\qquad \|u\|_m^2 \le c \sum_{|\delta|=m} \int_A |D^\delta u|^2 dx .$$

__The constant__ c __depends only on__ A.

It is evident that it suffices to prove (3.6) for $m=1$, and then, by induction, the general case follows. We have:

$$|c_0|^2 = \frac{1}{(2\pi)^r} \left|\int_A u\, dx\right|^2 = \frac{1}{(2\pi)^r} \frac{1}{r^2} \left|\int_A (Du)x\, dx\right|^2 \le$$

$$\le \frac{1}{(2\pi)^r} \frac{1}{r^2} \int_A |Du|^2 dx \int_A |x|^2 dx .^{(1)}$$

It follows that:

$$\|u\|_1^2 = \int_A |u|^2 dx + \int_A |Du|^2 dx \le \sum_{-\infty}^{+\infty}{}_{\kappa} |c_\kappa|^2 + p_1 \sum_{-\infty}^{+\infty}{}_{\kappa} |\kappa|^2 |c_\kappa|^2 =$$

$$= |c_0|^2 + (1+p_1) \sum_{\kappa \ne 0} |\kappa|^2 |c_\kappa|^2 \le$$

$$\le \left[\frac{1}{(2\pi)^r} \frac{1}{r^2} \int_A |x|^2 dx + \frac{1+p_1}{p_0} \right] \int_A |Du|^2 dx .$$

---

(1) Du denotes the $r \times n$ matrix $\dfrac{\partial u_\kappa}{\partial x_h}$ $(h=1,\dots,r;\ \kappa=1,\dots,n)$.

From (3.6) and (3.5), it follows that:

3.II. The norms $\| u \|_m^2$ and $\sum_{-\infty}^{+\infty}{}_k |k|^{2m} |c_k|^2$ are isomorphic in the space $\mathring{H}_m$.

Let us now consider, for any $u \in \mathring{H}_m (m > c)$, the operators $\mathcal{J}_{m,l}$ with domain $\mathring{H}_m$ and range in $\mathring{H}_l$ (with $l < m$) that associates with u the same u, but considered as a vector of $\mathring{H}_l$. It is obvious that $\mathcal{J}_{m,l}$ is a bounded linear transformation of $\mathring{H}_m$ into $\mathring{H}_l$. This transformation will be called the embedding of $\mathring{H}_m$ into $\mathring{H}_l$.

3.III. The embedding $\mathcal{J}_{m,l}$ of $\mathring{H}_m$ into $\mathring{H}_l$ $(l < m)$ is compact.

Let U be a bounded set of $\mathring{H}_m$. Then a positive constant L exists such that $\sum_{-\infty}^{+\infty}{}_k |k|^{2m} |c_k|^2 < L$ for any $u \in U$. It follows that $\sum_{-\infty}^{+\infty}{}_k |k|^{2l} |c_k|^2 < L$ and that the last series is uniformly convergent when u varies in U. This proves compactness of $\mathcal{J}_{m l}$ (2).

If T is a bounded closed domain of X^n, we shall say that T is regular if the Gauss-Green identity holds in T for any $u \in C^1(T)$

$$\int_{\partial T} u \nu_k \, d\sigma = \int_T \frac{\partial u}{\partial x_k} \, dx \qquad (k = 1, \dots, n),$$

$\nu \equiv (\nu_1, \dots, \nu_n)$ is the exterior unit normal vector to ∂T in any regular point of ∂T; $d\sigma$ is the measure of the hypersurface element on ∂A. The function u is said to belong to $D^m(T)$ $(m > 0)$ whenever it belongs to $C^{m-1}(T)$ and there exists a decomposition of T into a finite number of non-overlapping regular domains $T_1, \dots, T_s$ (3)

---

(2) We have been using the following lemma: If S is a separable Hilbert space and $\{v_k\}$ a complete orthonormal system in it, the subset U of S is compact if and only if the series $\sum_k |(u, v_k)|^2$ is uniformly bounded and uniformly convergent in U. (See [2]).

(3) By "decomposition of T in a finite number of non-overlapping domains $T_1, \dots, T_s$" we mean that $T = T_1 \cup \dots \cup T_s$ and $(T_i - \partial T_i) \cap (T_j - \partial T_j) = \phi$ for $i \neq j$.

such that $u \in C^m(T_i)(i=1,\dots,s)$. When $u \in D^m(T)$ we shall say that u has piece-wise continuous mth derivatives. It is not difficult to prove that $D^m(T) \subset H_m(T-\partial T)$. Moreover if we denote by $\mathring{D}^m(T)$ the subset of $D^m(T)$ consisting of those functions u such that $\operatorname{spt} u \subset T-\partial T$, then $\mathring{D}^m(T) \subset \mathring{H}_m(T-\partial T)$.

A domain A is said to be a properly regular domain if the following hypotheses are satisfied:

α) $\partial A = \partial \bar{A}$, and $\bar{A}$ is regular;

β) for any $x_0 \in \partial A$ there exists a neighborhood I of x_0 (open set containing x_0), such that the set $J = \bar{I} \cap \bar{A}$ is homeomorphic to the closed semiball $B^+ : y_n \geq 0$, $|y| \leq 1$ of the cartesian space Y^n. In this homeomorphism the set $\partial A \cap \bar{I}$ is mapped onto the set: $y_n = 0$, $y_1^2 + \dots + y_{n-1}^2 \leq 1$;

γ) the vector-valued function $y = y(x)$ which maps J homeomorphically onto B^+ has piece-wise continuous first derivatives [i.e. $y(x) \in D^1(J)$] and the jacobian matrix $\frac{\partial y}{\partial x}$ has a positive determinant which is bounded away from zero in the whole J.

It follows that the inverse function $x = x(y)$, which maps B^+ onto J, has properties analogous to those of $y = y(x)$.

3.IV. If A is properly regular, the embedding $\mathcal{J}_{m,\ell}$ of H_m into H_ℓ $(m > \ell)$ is compact (Rellich selection principle).

Let $I_1, \dots, I_q$ be a finite set of neighborhoods, like those mentioned in the condition β), that cover the boundary ∂A of A. Let $I = I_1 \cup \dots \cup I_q$ and suppose that the set $A - I$ be nonempty (otherwise the proof is simpler). The closed set $A - I$ is contained in A. Let I_0 be any open subset of A containing $A - I$. The sets $I_0, I_1, \dots, I_q$ form an open covering of $\bar{A}$. Let $\sum_{h=1}^{\nu} \varphi_h(x) = 1$ be a partition of unity with non-negative C^∞ functions such that $\operatorname{spt} \varphi_h(x)$ $(h = 1, \dots, \nu)$

is contained in one of the sets of the above considered covering. Let $u \in H_1(A)$ We have $u = \sum_{h=1}^{\nu} \varphi_h u$ and

$$\|\varphi_h u\|_1^2 = \int_A |\varphi_h u|^2 dx + \sum_{k=1}^{r} \int_A \left| u \frac{\partial \varphi_h}{\partial x_k} + \varphi_h \frac{\partial u}{\partial x_k} \right|^2 dx \leq$$

$$\leq \int_A |\varphi_h u|^2 dx + 2\int_A |u|^2 \sum_{k=1}^{r} \left|\frac{\partial \varphi_h}{\partial x_k}\right|^2 dx + 2\int_A |\varphi_h|^2 \sum_{k=1}^{r} \left|\frac{\partial u}{\partial x_k}\right|^2 dx \leq c\,\|u\|_1^2$$

Let $\operatorname{spt} \varphi_h \subset \overset{\circ}{I}_{s(h)}$. Then if U is a bounded set in $H_1(A)$, the set U_h of functions $\varphi_h u$ $(u \in U)$ is bounded in $H_1(I_{s(h)} \cap A)$. Suppose we have proved that, for U_h bounded, from any sequence $\{u_n\} \in U$ we can extract a subsequence $\{u_{n_k}\}$ such that $\{\varphi_h u_{n_k}\}$ is convergent in $H_0(I_{s(h)} \cap A)$. Then we can obviously suppose that $\{\varphi_h u_{n_k}\}$ is convergent for any h. Hence $u_{n_k} = \sum_{h=1}^{\nu} \varphi_h u_{n_k}$ will be convergent in $H_0(A)$. This proves the theorem for $m = 1$, $l = 0$. The extension to general m and l $(m > l)$ is trivial.

Compactness of U_h in the space $H_0(I_{s(h)} \cap A)$ when $s(h) = 0$ follows from 3.III since in this case $U_h \subset \overset{\circ}{H}_1(I_0)$. Let $s(h)$ be greater than zero. Set, for simplicity, $I_{s(h)} = I$. For any $u \in H_1(I \cap A)$ set $\tilde{u}(y) = u[x(y)]$. We have almost everywhere on B^+:

$$\frac{\partial \tilde{u}}{\partial y_j} = \frac{\partial u}{\partial x_h} \frac{\partial x_h}{\partial y_j}.$$

It follows that:

$$\int_{B^+} |\tilde{u}|^2 dy + \sum_{j=1}^{r} \int_{B^+} \left|\frac{\partial \tilde{u}}{\partial y_j}\right|^2 dy = \int_J |u|^2 \left|\frac{\partial y}{\partial x}\right| dx +$$

$$+ \sum_{j=1}^{m} \int_J \frac{\partial x_h}{\partial y_j} \frac{\partial x_k}{\partial y_j} \frac{\partial u}{\partial x_h} \frac{\partial u}{\partial x_k} \left|\frac{\partial y}{\partial x}\right| dx \leq c_1 \|u\|_{1,J}^2$$

where $\| \quad \|_{1,J}$ has an obvious meaning.

Let us now assume $\tilde{u}(y) = \varphi_h[x(y)]u[x(y)]$ with $u \in U$. Let us define in the closed ball B : $|y| \leq 1$ the following function:

$$\tilde{u}^*(y) \begin{cases} = \tilde{u}(y_1,\dots,y_{n-1}, y_n) \text{for } y_n \geq 0 \\ \\ = \tilde{u}(y_1,\dots,y_{n-1}, -y_n) \text{for } y_n < 0. \end{cases}$$

The function $\tilde{u}^*(y)$ belongs to $\overset{\circ}{H}_1(B)$. We have $\|\tilde{u}^*\|^2_{1,B} = 2\|\tilde{u}\|^2_{1,B^+} \leq$ $\leq 2c_1 c \|u\|^2_1$. Then we can extract from any sequence $\{u_m\} \in U$ a subsequence $\{u_{m_k}\}$ such that $\{\tilde{u}^*_{m_k}\}$ is convergent in $H_0(B)$ (theorem 3.III). This means that $\{\tilde{u}_{m_k}\}$ is convergent in $H_0(B^+)$ and therefore that $\{\varphi_h u_{m_k}\}$ is convergent in $H_0(I \cap A)$.

Bibliography of Lecture 3

[1] R.COURANT-D.HILBERT - Methoden der Mathematischen Physik - vol.II Springer, Berlin.

[2] G.FICHERA - see [1] in lecture 1.

[3] G.FICHERA - see [2] in lecture 2.

[4] C.MORREY - Multiple integral problems in the calculus of variations and related topics - Univ. of Calif. Publ.in Math.new ser.1943.

[5] S.L.SOBOLEV - Applications of Functional Analysis to Mathematical Physics - Leningrad, 1950.

Lecture 4

The trace operator. Sobolev and Ehrling lemmas.

Let A be a properly regular domain of X^r. Let τ denote the linear transformation which to any $u \in C^m(\bar{A})$ associates its boundary values τu on ∂A. Set for $m \geq 1$:

$$\| \tau u \|^2_{m-1, \partial A} = \sum_{|k|=0}^{m-1} \int_{\partial A} | D^k u |^2 d\sigma ,$$

we want to prove that:

$$\| \tau u \|^2_{m-1, \partial A} \leq c \| u \|^2_m , \tag{4.1}$$

where c is a constant depending only on A. It is enough to prove (4.1) in the case $m = 1$. For this, let us consider the open covering $I_0, I_1, \dots, I_q$, already introduced in the proof of theor.3.IV, and the corresponding partition of unity $\sum_{h=1}^{\nu} \varphi_h(x) = 1$. Let h be such that $\operatorname{spt} \varphi_h \subset I_{s(h)}$ with $s(h) > 0$, then (4.1) will follow from the inequalities:

$$\int_{\partial A} |\varphi_h u|^2 d\sigma \leq c \left(\int_A |\varphi_h u|^2 dx + \sum_{k=1}^{r} \int_A \left| \frac{\partial \varphi_h u}{\partial x_k} \right|^2 dx \right). \tag{4.2}$$

As in lecture 3 we set $I = I_{s(h)}$ and consider the homeomorphism that maps $J = \bar{I} \cap \bar{A}$ onto B^+. We suppose that B^+ is the semiball of the cartesian (y, t)-space defined by $t \geq 0$, $y_1^2 + \dots + y_{r-1}^2 + t^2 \leq 1$. Let D be the $(r-1)$-ball: $t = 0$, $|y|^2 = \sum_{k=1}^{r-1} y_k^2 \leq 1$, i.e. the image of $\bar{I} \cap \partial A$ under the homeomorphism of J onto B^+. Let $\tilde{u}(y, t)$

be the function $\varphi_h(x)u(x)$ when considered in B^+, by means of the above mentioned homeomorphism. We assume $\tilde{u}(t,y)\equiv 0$ for $t>0$, $t^2+|y|^2>1$. We have for any $t>0$:

$$\tilde{u}(y,0) = \tilde{u}(y,t) + \int_t^0 \tilde{u}_s(y,s)\,ds.$$

Hence,

$$\int_D |\tilde{u}(y,0)|^2 dy \leq 2\int_D |\tilde{u}(y,t)|^2 dy + 2\int_0^1\int_D |\tilde{u}_t(y,t)|^2 dy\,dt.$$

It follows that

$$\int_D |\tilde{u}(y,0)|^2 dy \leq 2\iint_{B^+} |\tilde{u}(y,t)|^2 dy\,dt + 2\iint_{B^+} |D\tilde{u}|^2 dy\,dt.$$

From this, returning to the $x_1,\dots,x_n$ coordinates, (4.2) follows; (4.1) is proven.

From (4.1) it follows that the operator τ can be continuously extended to the whole space $H_m(A)$. Then for any u a set of vectors $\tau u \equiv \{\varphi^p\}$ $(0\leq|p|\leq m-1)$ is determined such that $\varphi^p\in\mathcal{L}^2(\partial A)$ and $\varphi^p\equiv D^p u$ if $u\in C^m(\bar{A})$. τ will be called the <u>trace-operator</u> and φ^p will be considered the boundary values of $D^p u$ in a generalised sense. It is obvious that $u\in \mathring{H}_m(A)$ implies $\tau u = 0$, i.e. any function of $\mathring{H}_m(A)$ vanishes (in a generalised sense) on ∂A together with any derivative of order $\leq m-1$.

4.I. <u>If</u> $u\in C^m(\bar{A})\cap\mathring{H}_m(A)$, <u>then</u> $D^p u = 0$ <u>on</u> ∂A $(0\leq|p|\leq m-1)$ <u>in the classical sense</u>.

The proof is obvious.

If u and v are two functions of $C^m(\bar{A})$ and $0<|p|\leq m$, the Gauss-Green formula holds:

$$(4.3)\qquad \int_A u D^p v\,dx + (-1)^{|p|}\int_A (D^p u)v\,dx = \int_{\partial A} M(u,v)\,d\sigma$$

where $M(u,v)$ is a bilinear differential operator of order $|p|-1$ in u and v respectively. From (4.1) it follows that (4.3) still holds if u and v are functions of $H_m(A)$, provided the boundary values of $D^q u$ and $D^q v$, $0 \leq q < p$, are understood in the above introduced generalised sense.

Let x^0 be any point of X^r and Σ the unit sphere $|x| = 1$. Let Γ be a set of positive measure on the unit sphere Σ and R a positive number. The set $C_{x^0}(\Gamma, R)$ of all x such that

$$\frac{x - x^0}{|x - x^0|} \in \Gamma, \qquad 0 < |x - x^0| \leq R$$

will be called a "cone". x^0 is the vertex of the cone.

The domain A is said to satisfy the "cone hypothesis" if for every point x^0 of $\bar{A}$ there exists a cone $C_{x^0}(\Gamma_{x^0}, R)$ of vertex x^0 which is contained in A, and is such that R is independent of x^0 and Γ_{x^0} is congruent to a fixed set Γ on Σ.

We let the reader prove that a properly regular domain satisfies the cone hypothesis.

4.II. (Sobolev lemma). If A is a bounded domain of X^r satisfying the cone hypothesis, the functions of any space $H_m(A)$, with $m > \frac{r}{2}$, are continuous in $\bar{A}$ (i.e. $H_m(A) \subset C^0(\bar{A})$) and

$$\max_{\bar{A}} |u| \leq c \, \|u\|_m \tag{4.4}$$

with c depending only on A.

It is enough to prove the inequality (4.4) in order to have the proof of the entire lemma.

Let x^0 be any point of A. We denote by $\varphi(x)$ a C^∞ function satisfying the following conditions:

$$\varphi(x) \begin{cases} = 1 & \text{for } |x - x^0| < \frac{R}{2} \\ = 0 & \text{for } |x - x^0| \geq R. \end{cases}$$

We have:

$$u(x^0) = -\int_0^R \frac{\partial \varphi u}{\partial \rho}\, d\rho = \frac{(-1)^m}{(m-1)!}\int_0^R \rho^{m-1}\frac{\partial^m \varphi u}{\partial \rho^m}\, d\rho$$

Integrating over Γ_{x^0} on the unit sphere, we obtain:

$$|\operatorname{meas}\Gamma\, u(x^0)| = \left|\frac{1}{(m-1)!}\int_{C_{x^0}(\Gamma_{x^0},R)} \rho^{m-r}\frac{\partial^m \varphi u}{\partial \rho^m}\, dx\right| \leq$$

$$\leq \frac{1}{(m-1)!}\left(\int_{C_{x^0}(\Gamma_{x^0},R)} \left|\frac{\partial^m}{\partial \rho^m}\varphi u\right|^2 dx\right)^{\frac{1}{2}}\left(\int_{C_{x^0}(\Gamma,R)} \rho^{2(m-r)}\, dx\right)^{\frac{1}{2}}.$$

The last integral is finite since $2m > r$. Then:

$$|u(x^0)| \leq \frac{1}{(m-1)!\operatorname{meas}\Gamma}\left(\int_{C_{x^0}(\Gamma,R)} \rho^{2(m-r)}\, dx\right)^{\frac{1}{2}}\|\varphi u\|_m .$$

Thus, (4.4) has been proved.

Let B_1, B_2, B_3 be three complex Banach spaces and assume that, as vector spaces, they satisfy the following inclusion conditions:

$$B_1 \subset B_2 \subset B_3 .$$

Let us assume that the embedding $\mathfrak{J}_{12}$ of B_1 into B_2 is compact and the imbedding $\mathfrak{J}_{23}$ of B_2 into B_3 is continuous. For any given $\varepsilon > 0$ there exists a positive constant $c(\varepsilon)$ such that for any $u \in B_1$:

(4.5) $$\|u\|_{B_2} \leq \varepsilon\|u\|_{B_1} + c(\varepsilon)\|u\|_{B_3} .$$

Let us suppose that (4.5) is not true for some $\varepsilon > 0$. Then for any positive integer n, there must exist some u_n such that:

$$\|u_n\|_{B_2} > \varepsilon \|u_n\|_{B_1} + n \|u_n\|_{B_3} .$$

Set $v_n = \frac{u_n}{\|u_n\|_{B_1}}$. We have:

(4.6) $\|v_n\|_{B_2} > \varepsilon + n \|v_n\|_{B_3}$, (4.7) $\|v_n\|_{B_1} = 1$.

It follows from (4.7) that $\|v_n\|_{B_2}$ is uniformly bounded with respect to n. Then, from (4.6), it follows that $\lim_{n\to\infty} \|v_n\|_{B_3} = 0$. On the other hand, since the sequence $\{v_n\}$ is compact in B_2, we can extract a subsequence $\{v_{n_k}\}$ converging in B_2 to some v. Because of the continuity of $\mathcal{J}_{23}$, $\{v_{n_k}\}$ converges to v in B_3. Then $v = 0$. But this contradicts $\|v_{n_k}\|_{B_2} > \varepsilon$ that follows from (4.6).

4.III. (First Ehrling lemma). Let A be any bounded domain of X^r. For any $\varepsilon > 0$ there exists a positive constant $c(\varepsilon)$ (depending only on ε, A and m) such that for any $u \in \mathring{H}_m(A)$ the following inequality holds:

(4.8) $$\|u\|_{m-1} \leq \varepsilon \|u\|_m + c(\varepsilon) \|u\|_0 .$$

The lemma follows as a particular case of (4.5), by using theor. 3.III.

4.IV. (Second Ehrling lemma). Let A be any properly regular domain of X^r. For any $\varepsilon > 0$ there exists a positive constant $c(\varepsilon)$ (depending only on ε, A and m) such that for any $u \in H_m(A)$ the inequality (4.8) holds.

The lemma follows as a particular case of (4.5), by using theor. 3.IV.

Bibliography of Lecture 4

[1] G. EHRLING - On a type of eigenvalue problem for certain elliptic differential operators - Math. Scandinavica, vol.2, 1954.

[2] G. FICHERA - Sull'esistenza e sul calcolo delle soluzioni dei problemi al contorno, relativi all'equilibrio di un corpo elastico - Annali Scuola Norm.Sup.Pisa, s.III - vol.IV, 1950.

[3] G. FICHERA - see [2] of lecture 1.

[4] L. NIRENBERG - Remarks on Strongly Elliptic Partial Differential Equations - Comm. on pure and appl. math. vol.8, New York, 1955.

[5] S.L. SOBOLEV - On a theorem of functional analysis - Mat.Sbornik N.S. 4, 1938.

Lecture 5

Elliptic linear systems. Interior regularity.

Let A be a domain of X^n. Suppose that the $m\times m$ complex matrices $a_s(x)$ $(0 \leq |s| \leq \nu)$ are defined in A. Consider the linear matrix differential operator $Lu \equiv a_s(x) D^s u$. This operator is said to be an elliptic operator in A if, for any real non zero n-vector ξ, the following condition is satisfied:

$$\det \sum_{|s|=\nu} a_s(x)\xi^s \neq 0$$

at every point $x \in A$ (1).

Examples: i) If $m=1$ and $\nu=1$, we have the linear operator $L \equiv a_1(x)\frac{\partial}{\partial x_1} + \cdots + a_n(x)\frac{\partial}{\partial x_n} + a_0(x)$. In this case L may be elliptic if and only if $n \leq 2$. In the case $n=1$, ellipticity for the operator $L \equiv a_1(x)\frac{d}{dx} + a_0(x)$ means $a_1(x) \neq 0$ at any point of the interval A of the real axis. In the case $n=2$, the operator $L \equiv a_1(x)\frac{\partial}{\partial x_1} + a_2(x)\frac{\partial}{\partial x_2} + a_0(x)$ is elliptic only when $\Im m\, a_1(x)\bar{a}_2(x) \neq 0$. For instance, the Cauchy-Riemann (or Wirtinger) operator $\frac{\partial}{\partial x} + i\frac{\partial}{\partial y}$ $\left(\frac{\partial}{\partial x} - i\frac{\partial}{\partial y}\right)$ is elliptic.

---

(1) A more general definition of elliptic operator has been given by Douglis & Nirenberg (see [4], [10]).

ii) If $n=1$, $\nu=2$, the linear operator is as follows:
$L = a_{ij}\dfrac{\partial^2}{\partial x_i \partial x_j} + b_i \dfrac{\partial}{\partial x_i} + c$. If the coefficients $a_{ij}(x)$ are real, L is elliptic if and only if the quadratic form $a_{ij}(x)\xi_i\xi_k$ is definite (positive or negative).

iii) Assume $n=3$ and consider the differential operator of classical elastostatics: $L = u_{i/hh} + \sigma u_{h/ih}$. It is easily seen that this operator is elliptic for any $\sigma \neq -1$.

In lecture 1 we considered the weak formulation of a general BVP. The problem consisted in proving the existence of a vector u of the B -space S_u satisfying the system of integral equations:

(5.1) $$\langle u, L^* v\rangle = \langle f, v\rangle$$

for any $v \in V$ and for a given $f \in S_f$. Let us suppose that $B_u = B_f = \mathcal{L}^2(A)$. Then (5.1) can be written as follows:

(5.2) $$\int_A u\, L^*(v)\, dx = \int_A f v\, dx .$$

If the "boundary operators" M_h are actual boundary operators (differential or integral), then V contains the linear variety $\overset{\circ}{C}{}^{\nu}(\bar{A})$. No matter what the boundary conditions are, a weak solution of any BVP (with datum and unknown function in $\mathcal{L}^2(A)$) must satisfy the system of integral equations (5.2).

We have the following <u>interior regularity theorem</u> for elliptic systems:

5.I. <u>If A is any domain of X^n , L is an elliptic operator in A and $a_s(x) \in C^\infty(A)$, $f(x) \in C^\infty(A)$, then any function satisfying (5.2) for any $v \in \overset{\circ}{C}{}^\infty$ with $\operatorname{spt} v \in A$ belongs to $C^\infty(A)$.</u>

Set $L_0(x,D) = \sum_{|s|=\nu} a_s(x) D^s$, $L^*(x,D) = \sum_{|s|=0}^{\nu} (-1)^{|s|} D^s \overline{a_s(x)}$,

$$L_0^*(x,D) = (-1)^\nu \sum_{|s|=\nu} \bar{a}_s(x) D^s .$$

We can suppose - without any loss in generality - that $\bar{A} \subset Q$ $(Q: |x_k| < \pi)$. Let x^0 be any point of A. Let $\Gamma_{2\delta}$ be a closed ball, with center x^0 and radius 2δ, contained in A. Let Γ_δ be the concentric ball of radius δ. We denote by $\varphi(x)$ a C^∞ real function that is equal to 1 at any point of Γ_δ and vanishes identically outside $\Gamma_{2\delta}$. Let γ be an arbitrary constant complex n-vector. Set $v = \varphi(x) e^{ikx}\gamma$; k is a vector different from zero with integer components. We have:

$$L^*(x,D)v = \varphi(x) L_0^*(x, ik) e^{ikx}\gamma + \sum_{|s|=0}^{\nu-1} \Phi_s(x) k^s e^{ikx}\gamma .$$

The $n \times n$ matrices $\Phi_s(x)$ have their supports contained in $\Gamma_{2\delta}$. From (5.2) we obtain, because of the arbitrariness of γ:

$$L_0(x^0, ik)\int_A \varphi u e^{-ikx} dx = -\sum_{|s|=0}^{\nu-1} k^s \int_A \bar{\Phi}_s u e^{-ikx} dx +$$

$$+ \int_A \varphi f e^{-ikx} dx + \int_A [L_0(x^0, ik) - L_0(x, ik)] \varphi u e^{-ikx} dx .$$

Since $\det L_0(x^0, ik) \neq 0$ (L is elliptic!), we have:

$$\int_A \varphi u e^{-ikx} dx = -\sum_{|s|=0}^{\nu-1} k^s [L_0(x^0, ik)]^{-1} \int_A \bar{\Phi}_s(x) u e^{-ikx} dx +$$

$$+ [L_0(x^0, ik)]^{-1} \int_A \varphi f e^{-ikx} dx +$$

$$+ \left[L_0(x^0, i\kappa)\right]^{-1} \int_A \left[L_0(x^0, i\kappa) - L_0(x, i\kappa)\right] \varphi u e^{-i\kappa x} dx.$$

Set:

$$L_0(x^0, i\kappa) - L_0(x, i\kappa) = \sum_{|s|=\nu} \alpha_s(x) \kappa^s \quad ; \quad \varphi u = U.$$

Let h **be a non vanishing constant real** n**-vector. We have:**

$$\frac{e^{i\kappa h} - 1}{|h|} \int_A U e^{-i\kappa x} dx =$$

(5.3)
$$= -\sum_{|s|=0}^{\nu-1} \kappa^s \left[L_0(x^0, i\kappa)\right]^{-1} \frac{e^{i\kappa h} - 1}{|h|} \int_A \bar{\Phi}_s(x) u e^{-i\kappa x} dx +$$
$$+ \left[L_0(x^0, i\kappa)\right]^{-1} \frac{e^{i\kappa h} - 1}{|h|} \int_A \varphi f e^{-i\kappa x} dx +$$
$$+ \sum_{|s|=\nu} \kappa^s \left[L_0(x^0, i\kappa)\right]^{-1} \frac{e^{i\kappa h} - 1}{|h|} \int_A \alpha_s(x) U(x) e^{-i\kappa x} dx.$$

Let $|h|$ **be less than the distance between** $\Gamma_{2\delta}$ **and** ∂A. **We have:**

$$\frac{e^{i\kappa h} - 1}{|h|} \int_A \alpha_s(x) U(x) e^{-i\kappa x} dx = \int_Q \frac{\alpha_s(x+h) U(x+h) - \alpha_s(x) U(x)}{|h|} e^{-i\kappa x} dx =$$

$$= \int_Q \alpha_s(x+h) \frac{U(x+h) - U(x)}{|h|} e^{-i\kappa x} dx + \int_Q \frac{\alpha_s(x+h) - \alpha_s(x)}{|h|} U(x) e^{-i\kappa x} dx.$$

Let us denote by c_k the Fourier coefficient of U and suppose that $u \in H_\ell(\Gamma_{2\delta})$. From (5.3), we obtain (using the O Landau symbol):

$$\sum_{|k|<m} |k|^{2\ell} \left| \frac{e^{ikh}-1}{|h|} \right|^2 |c_k|^2 = O(\|u\|^2_{\ell,\Gamma_{2\delta}}) +$$

$$+ O\left(\sum_{-\infty}^{+\infty}{}_k |k|^{2(1-\nu+\ell)} \left| \int_Q \varphi f e^{-ikx} dx \right|^2 \right) +$$

$$+ \sum_{|s|=\nu} \left\{ O\left(\| \alpha_s(x+h) \frac{U(x+h)-U(x)}{|h|} \|^2_\ell \right) + O\left(\| \frac{\alpha_s(x+h)-\alpha_s(x)}{|h|} U(x) \|^2_\ell \right) \right\}.$$

Since the $(\ell+1)$-th derivatives of $\alpha_s(x)$ are bounded, ($\alpha_s(x) \in C^\infty$!) we have:

$$O\left(\| \frac{\alpha_s(x+h)-\alpha_s(x)}{|h|} U(x) \|^2_\ell \right) = O\left(\| u(x) \|^2_{\ell,\Gamma_{2\delta}} \right).$$

Let σ_ε be such that (for any $|s| = \nu$):

$$\max_{|x+h-x^0| \leq 2\sigma_\varepsilon} |\alpha_s(x+h)| < \varepsilon$$

and assume $2\delta < \sigma_\varepsilon$ and $|h| < \sigma_\varepsilon$; ε is chosen such that:

$$O\left(\| \alpha_s(x+h) \frac{U(x+h)-U(x)}{|h|} \|^2_\ell \right) \leq \frac{1}{2(2\pi)^2} \sum_{-\infty}^{+\infty}{}_k |k|^{2\ell} \left| \int_Q \frac{U(x+h)-U(x)}{|h|} e^{-ikx} dx \right|^2 +$$

(5.4)

$$+ c \sum_{-\infty}^{+\infty}{}_k |k|^{2\ell-2} \left| \int_Q \frac{U(x+h)-U(x)}{|h|} e^{-ikx} dx \right|^2 ,$$

where c is a positive constant.

This choice is possible since we have:

$$\left\| \alpha_s(x+h) \frac{U(x+h)-U(x)}{|h|} \right\|_\ell^2 \leq c_1 \sum_{|\kappa|=\ell} \int_Q \left| \alpha_s(x+h) D^\kappa \frac{U(x+h)-U(x)}{|h|} \right|^2 dx +$$

$$+ c_2 \sum_{|\kappa|<\ell} \int_Q \left| \beta_{s\kappa}(x+h) D^\kappa \frac{U(x+h)-U(x)}{|h|} \right|^2 dx.$$

where c_1 and c_2 are positive constants and $|\beta_{s\kappa}(x+h)|$ are bounded functions. Then, by using (3.5), the inequality (5.4) follows.

From (5.4), we get:

$$O\left(\left\| \alpha_s(x+h) \frac{U(x+h)-U(x)}{|h|} \right\|_\ell^2 \right) \leq \frac{1}{2} \sum_{-\infty}^{+\infty}{}_\kappa |\kappa|^{2\ell} \left| \frac{e^{i\kappa h}-1}{|h|} \right|^2 |c_\kappa|^2 + O\left(\|u\|^2_{\ell,\Gamma_{2\sigma}} \right).$$

Then, from (5.3) it follows that:

$$\frac{1}{2} \sum_{-\infty}^{+\infty}{}_\kappa |\kappa|^{2\ell} \left| \frac{e^{i\kappa h}-1}{|h|} \right|^2 |c_\kappa|^2 = O\left(\|u(x)\|^2_{\ell,\Gamma_{2\sigma}} \right) +$$

$$+ O\left(\sum_{-\infty}^{+\infty}{}_\kappa |\kappa|^{2(1-\nu+\ell)} \left| \int_Q \varphi f e^{-i\kappa x} dx \right|^2 \right).$$

By considering the following successive choices of h, $h \equiv (t,0,\dots,0)$, $h \equiv (0,t,\dots,0)$, ... $h \equiv (0,0,\dots,t)$ and making $t \to 0$ we obtain:

$$\sum_{-\infty}^{+\infty}{}_{\kappa} |\kappa|^{2(\ell+1)} |c_\kappa|^2 = O\left(\|u(x)\|^2_{\ell,\Gamma_{2\delta}}\right) + O\left(\sum_{-\infty}^{+\infty}{}_{\kappa} |\kappa|^{2(1-\nu+\ell)} \left|\int_Q \varphi f e^{-i\kappa x} dx\right|^2\right).$$

We have so far proven that if $u \in H_\ell(\Gamma_{2\delta})$ then $\varphi u \in H_{\ell+1}(\Gamma_{2\delta})$ and consequently $u \in H_{\ell+1}(\Gamma_\delta)$. Since $u \in \mathcal{L}^2(A) = H_0(A)$, we have proved that for any $x^0 \in A$ and any positive integer ℓ there exists a circular neighborhood I of x^0 such that u belongs to $H_\ell(I)$. From the Sobolev lemma, it follows that u is C^∞ in A, and therefore u is a solution of the differential equation $Lu = f$ in the classical sense.

If we look back carefully at the proof of the theorem, we see that the arguments we used have proven the following more general theorem:

5.II. A is any domain in the space X^r; $L \equiv a_s(x) D^s$ is an elliptic matrix-differential operator such that $a_s(x) \in C^{\ell+|s|}(A)$ ($\ell \geq 0$); f a function belonging to $H_{1-\nu+\ell}(A)$ (to $H_0(A)$ if $1-\nu+\ell < 0$). For any $x^0 \in A$ there exists a δ such that

$$\|u\|^2_{\ell+1,\Gamma_\delta} \leq c\left(\|f\|^2_{1-\nu+\ell,\Gamma_{2\delta}} + \|u\|^2_{\ell,\Gamma_{2\delta}}\right) \qquad (1)$$

for any $\mathcal{L}^2$-weak solution u of the equation $Lu = f$. The constant c depends only on L and on x^0.

From this theorem it follows, by using the Sobolev lemma, that the interior regularity theorem holds when $a_s(x) \in C^h(A)$, with $h > \frac{r}{2} + \nu - 1 + |s|$, and $f(x) \in H_\kappa(A)$, with $\kappa > \frac{r}{2}$. This means that under these hypotheses any $\mathcal{L}^2$-weak solution of $Lu = f$ belongs to $C^\nu(A)$.

---

(1) Now and in the sequel read $\| \ \|^2_0$ instead of $\| \ \|^2_m$ if $m < 0$.

Bibliography of Lecture 5.

[1] F.E. BROWDER - The Dirichlet problem for linear elliptic equations of arbitrary even order with variable coefficients - Proc.Nat.Acad.Sci.U.S.A.,vol.38, 1952.

[2] F.E. BROWDER - Assumption of boundary values and the Green's function in the Dirichlet problem for the general elliptic equation - Proc.Nat.Acad.Sci.USA,vol.39,1953.

[3] R. CACCIOPPOLI - Sui teoremi di esistenza di Riemann - Ann.Scuola Norm.Sup.Pisa, 1939.

[4] A. DOUGLIS-L. NIRENBERG - Interior estimates for elliptic systems of partial differential equations - Comm. Pure Appl. Mathem.,vol.8, 1955.

[5] K.O. FRIEDRICHS - On the differentiability of the solutions of linear elliptic differential equations - Comm. Pure Appl. Mathem.,vol.6, 1953.

[6] F. JOHN - The fundamental solution of linear elliptic differential equations with analytic coefficients - Comm. Pure Appl.Mathem., vol.3, 1950.

[7] F. JOHN - General properties of solutions of linear elliptic partial differential equations - Proc.Sump.Spectral Theory and Diff. Problems, Oklahoma College, 1951.

[8] F. JOHN - Derivatives of continuous weak solutions of linear elliptic equations - Comm. Pure Appl. Math., vol. 6, 1953.

[9] P. LAX - On Cauchy's problem for hyperbolic equations and the differentiability of solutions of elliptic equations - Comm. Pure Appl. Math., vol. 8, 1955.

[10] L. NIRENBERG - see [4] of lecture 4.

[11] H. WEYL - The method of orthogonal projection in potential theory - Duke Math. Journal, vol. 7, 1940.

Lecture 6

Existence of local solutions for elliptic systems.

The aim of this lecture is to prove that for any point x^0 of the domain A, where the elliptic operator L, introduced in lecture 5, is considered, there exists some neighborhood in which a solution u of the differential system $Lu = f$ exists.(1)

We first prove some lemmas.

6.I. Let the coefficients $a_s(x)$ of the elliptic operator L satisfy the conditions stated in theor. 5.II. Let A_0 be a bounded domain such that $\bar{A}_0 \subset A$. There exists a positive constant c_0, depending only on L and on A_0, such that, for any $v \in \overset{\circ}{C}{}^{\infty}(A_0)$ the following inequality holds:

$$(6.1) \qquad \|v\|^2_{l+1,A_0} \leq c_0 \left(\|Lv\|^2_{1-\nu+l,A_0} + \|v\|^2_{l,A_0} \right).$$

We can cover A_0 by a finite set of closed balls $\Gamma_{\delta_1},\dots,\Gamma_{\delta_q}$, such that, for any Γ_{δ_i} $(i=1,\dots,q)$, the following inequality holds (we assume $v \equiv 0$ in $X^r - A$):

$$\|v\|^2_{l+1,\Gamma_{\delta_i}} \leq c_i \left(\|Lv\|^2_{1-\nu+l,\Gamma_{2\delta_i}} + \|v\|^2_{l,\Gamma_{2\delta_i}} \right)$$

(theor. 5.II). Since:

$$\|v\|^2_{l+1,A_0} \leq \sum_{i=1}^{q} \|v\|^2_{l+1,\Gamma_{\delta_i}},$$

---

(1) Please note that the example of H.Lewy considered in Lecture 1 refers to an operator which is not elliptic.

$$\| L v \|^2_{1-\nu+\ell, \Gamma_{2\delta_i}} \leq \| L v \|^2_{1-\nu+\ell, A_0} , \quad \| v \|^2_{\ell, \Gamma_{2\delta_i}} \leq \| v \|^2_{\ell, A_0} ,$$

(6.1) follows with $c_0 = \sum_{i=1}^{4} c_i$.

Let us assume - for the sake of simplicity - from now on, $a_s(x) \in C^\infty(A)$. We leave it as an exercise to the reader to recognize the more general hypotheses, under which what we are going to say holds.

6.II. There exists only a finite number of linearly independent solutions of the homogeneous equation $Lv = 0$ belonging to $\overset{\circ}{C}{}^\infty(A_0)$. Let ℓ be any non-negative integer. For any solution v of $Lv = 0$, belonging to $\overset{\circ}{C}{}^\infty(A_0)$, the following inequality holds:

$$\| v \|^2_{\ell+1, A_0} \leq c_0 \| v \|^2_{\ell, A_0} . \tag{6.2}$$

Let V_0 be the linear variety of solutions of $Lv = 0$ belonging to $\overset{\circ}{C}{}^\infty(A_0)$. From (6.2) it follows that if we consider V_0 as a subvariety of $\overset{\circ}{H}_\ell(A_0)$, any bounded subset of V_0 is compact (theor. 3.III). Then V_0 is finite-dimensional.

6.III. For any $x^0 \in A$ a positive number ϱ_0 exists such that in the open ball Γ_{ϱ_0} , of center x^0 and radius ϱ_0, the homogeneous equation $Lv = 0$ has only the trivial solution $v \equiv 0$ belonging to $\overset{\circ}{C}{}^\infty(\Gamma_{\varrho_0})$. Let δ be such that $\Gamma_\delta \subset A$ (Γ_δ is the open ball of center x^0 and radius δ). Let us denote by $v_1, \ldots, v_m$ a complete system of linearly independent solutions of $Lv = 0$, belonging to $\overset{\circ}{C}{}^\infty(\Gamma_\delta)$. For any positive $\varrho < \delta$, we denote by $\Gamma_{\delta,\varrho}$ the open set $\varrho < |x - x^0| < \delta$. Let us consider the Gramian determinant:

$$\gamma(\varrho) = \begin{vmatrix} \int_{\Gamma_{\delta,\varrho}} |v_1|^2 dx & \cdots & \int_{\Gamma_{\delta,\varrho}} v_1 v_m \, dx \\ \cdot & \cdots & \cdot \\ \int_{\Gamma_{\delta,\varrho}} v_m v_1 \, dx & \cdots & \int_{\Gamma_{\delta,\varrho}} |v_m|^2 dx \end{vmatrix} .$$

We can choose ϱ_0 such that $\gamma(\varrho_0) > 0$. If v belongs to $\overset{\circ}{C}{}^{\infty}(\Gamma_{\varrho_0})$ and is a solution of $Lv = 0$, it is expressed in Γ_δ as a linear combination of $v_1, \dots, v_m$, i.e. $v = \sum_{k=1}^{m} a_k v_k$. But since $v_1, \dots, v_m$ are linearly independent in $\Gamma_{\delta,\varrho_0}$, this implies $a_1 = \dots = a_m = 0$. Then $v \equiv 0$ in Γ_{ϱ_0}.

6.IV. <u>Assign to</u> x^0 <u>and</u> ϱ^0 <u>the same meaning as in theor.6.III. In the open ball</u> Γ_ϱ <u>with</u> $0 < \varrho < \varrho_0$, <u>the following inequality holds:</u>

$$\| v \|^2_{l+1, \Gamma_\varrho} \leq c \| L v \|^2_{1-\nu+l, \Gamma_\varrho} \tag{6.3}$$

<u>for any</u> $v \in \overset{\circ}{C}{}^{\infty}(\Gamma_\varrho)$.

In fact, suppose (6.3) were not true. Then there is a sequence $\{v_k \in \overset{\circ}{C}{}^{\infty}(\Gamma_\varrho)\}$ such that:

$$\| v_k \|^2_{l+1, \Gamma_\varrho} = 1, \qquad \| L v_k \|^2_{1-\nu+l, \Gamma_\varrho} \leq \frac{1}{k}. \tag{6.4}$$

By theor. 3.III there is a subsequence, which we still denote by $\{v_k\}$, which converges in $\overset{\circ}{H}_l(\Gamma_\varrho)$. Furthermore $\{L v_k\}$ converges to zero in the space $\overset{\circ}{H}_{1-\nu+l}(\Gamma_\varrho)$. From the inequality (6.1), considered for $A_0 = \Gamma_\varrho$, it follows that $\{v_k\}$ converges in $\overset{\circ}{H}_{l+1}(\Gamma_\varrho)$ and, consequently, in $H_{l+1}(\Gamma_{\varrho_0})$. Let v be the limit of $\{v_k\}$ in $H_{l+1}(\Gamma_{\varrho_0})$. We have for any $w \in \overset{\circ}{C}{}^{\infty}(\Gamma_{\varrho_0})$:

$$\int_{\Gamma_{\varrho_0}} v_k L^* w \, dx = \int_{\Gamma_{\varrho_0}} (L v_k) w \, dx .$$

For $k \to \infty$, it follows that

$$\int_{\Gamma_{\varrho_0}} v L^* w \, dx = 0 .$$

Then $v \in C^{\infty}(\Gamma_{\rho_0})$ and since it vanishes identically in $\Gamma_{\rho_0, \rho}$, v is a solution of the equation $Lv = 0$ belonging to $\mathring{C}^{\infty}(\Gamma_{\rho_0})$. By the preceding lemma, we have $v \equiv 0$ in Γ_{ρ_0}. But this result contradicts the fact that $\| v \|^2_{\ell+1, \Gamma_\rho} = 1$, which follows from (6.4).

We are now in position to prove the existence theorem for the local solution of the differential system $Lu = f$.

6.V. Let $f \in C^{\infty}(A)$. For any $x^\circ \in A$ there exists a positive number ρ such that in the ball Γ_ρ, $|x^\circ - x| < \rho$, there exists some solution of the equation $Lu = f$.

Let ρ be such that for any $v \in \mathring{C}^{\infty}(\Gamma_\rho)$ inequality (6.3) holds with $\ell = \nu - 1$, and L replaced by L^*. Then we deduce the following inequality:

$$\| v \|^2_{0,\Gamma_\rho} \leq c \, \| L^* v \|^2_{0,\Gamma_\rho} .$$

It follows, from the existence principle of lecture 2, that there exists an $u \in \mathcal{L}^2(\Gamma_\rho)$ which satisfies the integral equations

$$\int_{\Gamma_\rho} u L^* v \, dx = \int_{\Gamma_\rho} f v \, dx$$

for any $v \in \mathring{C}^{\infty}(\Gamma_\rho)$. This proves the theorem.

In general, if A_0 is any bounded subdomain of A, a solution of $Lu = f$ exists in A_0 if and only if the inequality

$$\| v \|^2_{0,A_0} \leq c \, \| L^* v \|^2_{0,A_0} \tag{6.5}$$

holds for any $v \in \mathring{C}^{\infty}(A_0)$. This is a consequence of the existence principle of lecture 2. Of course, the solution is supposed to belong to $\mathcal{L}^2(A_0) \cap C^{\infty}(A_0)$.

The inequality (6.5) may fail to be true for some A_0 , as the examples constructed by Plis [3] prove. When (6.5) holds, we have the following "dual inequality":

$$\inf_{u_0 \in U_0} \| u + u_0 \|^2_{0,A_0} \leq c \| Lu \|^2_{0,A_0} ,$$

where $u \in \mathcal{L}^2(A_0) \cap C^\infty(A_0)$ and U_0 is the linear variety of all solutions of the homogeneous equation $Lu_0 = 0$ belonging to $\mathcal{L}^2(A_0)$.

Bibliography of Lecture 6

[1] J.PEETRE - Another approach to elliptic boundary problems - Comm. Pure and Appl. Math. vol.14, 1961.

[2] J.PEETRE - Elliptic Partial Differential Equations of Higher Order - Univ. of Maryland, Inst. Fluid Dyn. Appl. Math. series N° 40, 1962.

[3] A. PLIS - Non-uniqueness in Cauchy's Problem for Differential Equations of Elliptic type.- Journal of Math. and Mech. vol.9, 1960.

Lecture 7

Semiweak solutions of BVP for elliptic systems.

Let us consider an $n \times n$ matrix operator L of order $2m$ $(m \geq 1)$ with C^{∞} complex coefficients. For convenience we write the operator as follows:

$$L \equiv D^{p} a_{pq}(x) D^{q} \qquad 0 \leq |p| \leq m \quad , \ 0 \leq |q| \leq m .$$

The adjoint operator is:

$$L^{*} = (-1)^{|p|+|q|} D^{q} \bar{a}_{pq} D^{p}.$$

If L is supposed to be elliptic, the ellipticity condition is then written

(7.1) $$\det \sum_{|p|,|q|=m} a_{pq}(x) \xi^{p} \xi^{q} \neq 0 \quad (\xi = \text{real non zero } r\text{-vector}).$$

We suppose, for simplicity, that the coefficients $a_{pq}(x)$ of L are defined in the whole space (and belong to C^{∞}).
Let A be a bounded domain of X^{r}. To the operator L we associate the following bilinear form

$$B(u,v) = (-1)^{|p|} \int_A (a_{pq}(x) D^{q} u) D^{p} v \, dx = (-1)^{|p|} \int_A D^{q} u \, (\bar{a}_{pq}(x) D^{p} v) \, dx . ^{(1)}$$

---

(1) From now on we shall omit parentheses when writing $(a_{pq}(x) D^{q} u) D^{p} v$.

which is defined for every pair of vectors u, v belonging to $H_m(A)$.

Suppose we consider the following BVP:

(7.2) $\quad Lu = f$ in A, (7.3) $D^p u = 0 \quad (0 \leq |p| \leq m-1)$ on ∂A.

This is the classical "Dirichlet problem". We assume $f \in C^\infty$, and we first want to find a solution $u \in H_m(A)$. The boundary conditions (7.3) will be satisfied in a generalised sense, if A is properly regular and we require that u belong to $\mathring{H}_m(A)$ (see lecture 4). If $u \in \mathring{H}_m(A) \cap C^\infty(A)$ and $Lu = f$, then for any $v \in \dot{C}^m(A)$, the following identity will be satisfied:

(7.4) $$B(u,v) = \int_A f v \, dx.$$

It follows that (7.4) is satisfied by any $v \in \mathring{H}_m(A)$.

We are now led to the following generalised formulation of the Dirichlet problem:

I) Find a function u belonging to $\mathring{H}_m(A)$ such that (7.4) is satisfied for any $v \in \mathring{H}_m(A)$.

If $v \in \dot{C}^\infty(A)$, from (7.4) it follows that (5.2) is satisfied. For this reason we say that u is a semiweak solution of the Dirichlet problem. Theorem 5.I assures us that u satisfies (7.2) in the classical sense. We have already observed that (7.3) is satisfied in a generalised sense if A is properly regular. Under further assumption on ∂A we shall prove that the boundary conditions (7.3) are satisfied in the classical sense by a semiweak solution.

We want now to generalise problem I) in order to include BVP's different from the Dirichlet problem.

Let V be a subspace (closed linear subvariety) of the space $H_m(A)$ such that $\overset{\circ}{H}_m(A) \subset V \subset H_m(A)$.

II) <u>Find a solution belonging to</u> V <u>such that (7.4) is satisfied for any</u> $v \in V$.

If A is properly regular, we may consider a still more general problem.

Let us denote by $\mathcal{T}_{m-1}(\partial A)$ the Hilbert function space obtained by functional completion from the vectors τu $[u \in H_m(A)]$ by means of the norm $\|\tau u\|_{m-1,\partial A}$ (defined in lecture 4). We denote by $[\varphi, \psi]_{m-1}$ the scalar product of two vectors of $\mathcal{T}_{m-1}(\partial A)$. Let γ be a bounded linear transformation with domain V and range in $\mathcal{T}_{m-1}(\partial A)$. The following bilinear form $[\gamma u, \tau v]_{m-1}$ is continuous in $V \times V$. Set $\Phi(u,v) = B(u,v) + [\gamma u, \tau v]_{m-1}$.

III) <u>Find a function belonging to</u> V <u>such that the following equation is satisfied for any</u> $v \in V$:

(7.5) $$\Phi(u,v) = \int_A f v \, dx.$$

As an example, we want to show how some classical boundary value problems for a 2nd order elliptic equation can be included as particular cases of problem III). Let us consider the scalar differential operator with real coefficients:

$$Lu = \frac{\partial}{\partial x_i}\left[a_{ij}(x)\frac{\partial u}{\partial x_j}\right] + b_i(x)\frac{\partial u}{\partial x_i} + cu .$$

In this example we consider only real functions. We have:

$$B(u,v) = \int_A \left[-a_{ij}\frac{\partial v}{\partial x_i}\frac{\partial u}{\partial x_j} + v b_i \frac{\partial u}{\partial x_i} + cuv\right] dx .$$

Let $\partial A = \partial_1 A \cup \partial_2 A$, $\partial_1 A \cap \partial_2 A = \phi$. We do not exclude the possibility that $\partial_K A$ ($K=1$ or $K=2$) may be empty. However, if $\partial_K A$ is not empty, we suppose that $\partial_K A$ is made up of a finite number of smooth hypersurfaces. Assume that V is defined by the condition $\tau u = 0$ on $\partial_1 A$ for $u \in V$. We assume $V = \overset{\circ}{H}_1(A)$ if $\partial_2 A$ is empty, and $V = H_1(A)$ if $\partial_1 A$ is empty.

Let $\gamma u = h(x) u$, where $h(x)$ is a continuous scalar function defined on $\partial_2 A$. Equation (7.5) becomes:

$$(7.6) \qquad B(u,v) + \int_{\partial_2 A} h u v \, d\sigma = \int_A f v \, dx .$$

Assume that a solution u of equations (7.6) exists belonging to $C^1(A)$. [2] Then we have $u \in C^\infty(A)$, $Lu = f$ in A , $u = 0$ on $\partial_1 A$ and moreover :

$$B(u,v) - \int_{\partial_2 A} \left(a_{ij} \nu_i \frac{\partial u}{\partial x_j}\right) v \, d\sigma = \int_A f v \, dx ,$$

where $\nu \equiv (\nu_1 \dots \nu_r)$ is the interior unit normal vector to $\partial_2 A$. It follows that u is a solution of the BVP

$$Lu = f, \qquad u = 0 \quad \text{on } \partial_1 A \ , \quad a_{ij} \nu_i \frac{\partial u}{\partial x_j} + hu = 0 \quad \text{on } \partial_2 A .$$

This BVP includes, by proper choices of $\partial_1 A$, $\partial_2 A$, h many of the classical BVP for a second order elliptic equation.

---

(2) This hypothesis is assumed now only for the sake of simplicity, but in general it is not satisfied.

We wish now to give necessary and sufficient conditions for the existence and uniqueness of solutions of problem III). Of course this will include the case of problem II) $[\gamma \equiv 0]$ and problem I) $[\gamma \equiv 0, V = \overset{\circ}{H}_m(A)]$. However, when we are concerned with these particular cases, we need not impose on A the condition of being properly regular, but merely that it be a bounded domain of X^r.
Since the bilinear form is continuous in $V \times V$, there exists a linear continuous transformation of V into itself such that, whenever u and v are in V, one has:

$$\Phi(u,v) = (u, Tv)_m .$$

Hence (7.5) may be written:

$$(u, Tv)_m = \int_A f v \, dx . \tag{7.5}$$

From the existence principle of lecture 2, it follows that:

7.I. <u>A necessary and sufficient condition in order that, given an arbitrary</u> $f \in \mathcal{L}^2(A)$ <u>there exist a unique</u> $u \in V$ <u>satisfying (7.5) for any</u> $v \in V$, <u>is that one have</u>:

1°) $\|v\|_0^2 \leq c \|Tv\|_m^2 \quad (c > 0)$ <u>for any</u> $v \in V$.

2°) <u>The range</u> $T(V)$ <u>of</u> T <u>is dense in</u> V.

If 1°) and 2°) hold then the solution u of the problem satisfies the dual inequality:

$$\|u\|_m^2 \leq c \|f\|_0^2 . \tag{7.6}$$

Now define:

$$\Psi(u,v) = \frac{1}{2}\left[\Phi(u,v) + \overline{\Phi(v,u)}\right]$$

and consider the real quadratic form $\Psi(v,v) = \mathrm{Re}\,\Phi(v,v)$. We shall say that the quadratic form $\Psi(v,v)$ is <u>coercive</u> on the quadratic form $\|v\|_m^2$ in V, when there exists a constant $c_0 > 0$ such that:

(7.7) $$|\Psi(v,v)| \geq c_0 \|v\|_m^2 ,$$

for any $v \in V$.

7.II. <u>A sufficient condition in order that there exist a unique solution of problem III) is that $\Psi(v,v)$ be coercive on $\|v\|_m^2$ in V.</u>

Let T^* denote the adjoint of the transformation T which was introduced above, and define:

$$T_1 = \frac{1}{2}(T + T^*) \quad ; \quad T_2 = \frac{1}{2i}(T - T^*).$$

One obviously has:

$$\Psi(u,v) = (u, T_1 v)_m .$$

And hence if (7.7) holds, one obtains:

$$\|v\|_m^2 \leq \frac{1}{c_0}\left|(v, T_1 v)_m\right| \leq \frac{1}{c_0}\|v\|_m \|Tv\|_m .$$

Then:

(7.8) $$\|v\|_m \leq \frac{1}{c_0}\|Tv\|_m$$

which obviously implies condition 1°) of the preceding theorem. Condition 2°) is also satisfied, because, if $u \in V$ and satisfies $(u, Tv)_m = 0$ for any $v \in V$, then one has $(T_1 u, u)_m + i(T_2 u, u)_m = 0$. Since T_1 and T_2 are hermitian operators, one gets $(T_1 u, u)_m = 0$, which, in view of (7.7), means that $u = 0$.

Let us now prove, by means of an example, that the condition (7.7) used in theorem 7.II is not necessary, but is only sufficient. To this end we shall consider a very simple particular case in which, while 1°) and 2°) of theorem 7.I hold, (7.7) does not.

Let $r = 1$ and suppose that A coincides with the interval $(-1, 1)$ of the x-axis, $m = 1$ and V is $\overset{\circ}{H}_1(A)$. Let us put:

$$\Phi(u, v) = \int_{-1}^{1} |x|^{1/3} \frac{du}{dx} \frac{d\bar{v}}{dx} dx .$$

Equation (7.5) in the present case becomes:

$$\int_{-1}^{1} |x|^{1/3} \frac{du}{dx} \frac{d\bar{v}}{dx} dx = \int_{-1}^{1} f(x) \bar{v}(x) dx . \tag{7.9}$$

It is readily verified that, given $f \in \mathcal{L}^2$, there exists one and only one $u \in V$, satisfying (7.9) for any $v \in V$ and that it is defined by the equation:

$$u(x) = \int_{-1}^{x} |\xi|^{-1/3} F(\xi) d\xi - \left[\int_{-1}^{1} |\xi|^{-1/3} d\xi \right]^{-1} \left[\int_{-1}^{1} |\xi|^{-1/3} F(\xi) d\xi \right] \int_{-1}^{x} |\xi|^{-1/3} d\xi .$$

where we have put:

$$F(x) = \int_{-1}^{x} f(\xi) d\xi .$$

Therefore, the above-mentioned conditions 1°) and 2°) are certainly satisfied. However (7.7) is not. In the present case, (7.7) may be written:

$$(7.10) \qquad \int_{-1}^{1} |x|^{1/3} \left|\frac{dv}{dx}\right|^2 dx \geq c_0 \int_{-1}^{1} \left(\left|\frac{dv}{dx}\right|^2 + |v|^2 \right) dx .$$

Indeed from (7.10) it follows that any v which vanishes on the intervals $(-1,-\varepsilon)$ and $(\varepsilon, 1)$, where $0 < \varepsilon < 1$ and $\varepsilon^{1/3} < c_0$, must satisfy:

$$\int_{-\varepsilon}^{\varepsilon} \left(|x|^{1/3} - c_0\right) \left|\frac{dv}{dx}\right|^2 dx \geq 0,$$

and this is absurd if v is not identically zero.

Bibliography of Lecture 7

[1] G. FICHERA - Alcuni recenti sviluppi della teoria dei problemi al contorno per le equazioni alle derivate parziali - Atti Convegno Internazionale Equazioni alle Derivate Parziali, Trieste 1954 - Ediz. Cremonese, Roma 1955.

[2] G. FICHERA - see [2] of lecture 1.

[3] J.L. LIONS - Problèmes aux limites en theorie des distributions - Acta Mathem. v. 94, 1955.

[4] J.L. LIONS - Sur les problèmes aux limites du type dérivée oblique - Annals of Mathem. v. 64, 1956.

[5] M.I. VISIK - On general BVP for elliptic differential equations (in russian) Trudy Moscow Mat. Obsc. 1952.

Lecture 8

Regularity at the boundary : preliminary lemmas.

In the X^r space we shall denote by y_k the coordinate x_k, for $k = 1,\dots, r-1$ and by t the coordinate x_r. The point $(y_1,\dots,y_{r-1})$ of the $(r-1)$-dimensional space will be denoted by y and the point x of X^r by (y, t).
Let R be the open interval of X^r defined by: $-\pi < y_k < \pi$ $(k = 1,\dots, r-1)$, $0 < t < \pi$. We shall say that the function v (as usual we consider complex n-vector valued functions) belongs to $C^m_\sigma(R)$ (σ is any fixed positive number such that $0 < \sigma < \pi$) if and only if: i) $v \in C^m$; ii) $v \equiv 0$ when at least one of the following conditions is satisfied: $|y| > \sigma$, $t > \sigma$.
The closure of $C^m_\sigma(R)$ in the space $H_m(R)$ will be denoted by $H^\sigma_m(R)$.

Let $v(x)$ be a function of $C^m_\sigma(R)$. We consider the following function $v^*(x)$ defined in X^r by:

$$(8.1) \qquad v^*(y,t) \begin{cases} = v(y,t) & \text{for } t \geq 0 \\ = \sum_{j=1}^{m+1} \lambda_j v(y,-jt) & \text{for } t < 0. \end{cases}$$

The scalars λ_j are the solutions of the following albegraic system:

(8.2) $$\sum_{j=1}^{m+1} (-j)^k \lambda_j = 1 \qquad k = 0, \dots, m.$$

It is easily seen that the function $v^*(y,t)$ belongs to C^m and that its support is contained in the square $Q : |x_k| < \pi$.
By very simple computations it is easy to verify that:

(8.3) $$\| v^* \|^2_{i,Q} \leq c \| v \|^2_{i,R} \qquad i = 0, \dots, m$$

where c is a constant which does not depend on v.

If $u \in H^{\sigma}_m(R)$, the function u^*, defined by means of (8.1), (8.2), belongs to $\mathring{H}_m(Q)$. In fact if $\{v_k\} \in C^m_\sigma(R)$ and converges in $H_m(R)$ to u, then $\{v^*_k\}$ converges in the space $\mathring{H}_m(Q)$ towards u^*. Moreover we have:

(8.4) $$\| u^* \|^2_{m,Q} \leq c \| u \|^2_{m,R} .$$

We denote by D_y the symbolic y-differentiation vector $D_y \equiv (\frac{\partial}{\partial y_1}, \dots, \frac{\partial}{\partial y_{n-1}})$.

Let $u \in \mathcal{L}^2(A)$ (A being a bounded domain of X^n). We say that u has the $\mathcal{L}^2$-strong derivative $D^p u$ ($|p| = m$) whenever a sequence of $C^m(\bar{A})$-functions $\{v_k\}$ exists such that v_k converges in $\mathcal{L}^2(A)$ towards u and $\{D^p v_k\}$ converges in $\mathcal{L}^2(A)$ to some function ψ^p which we define to be the $\mathcal{L}^2$-strong derivative $D^p u$ of u. Hence:

(8.5) $$\int_A u \, D^p v \, dx = (-1)^{|p|} \int_A \psi^p v \, dx$$

for any $v \in \overset{\circ}{C}{}^{m}(A)$. It follows that ψ^{p} does not depend on $\{v_\kappa\}$. If Γ is a closed subset of A and u vanishes in $A-\Gamma$, then, if u has the $\mathcal{L}^2$-strong derivative $D^p u$, this derivative vanishes in $A-\Gamma$. In that case (8.5) holds for any $v \in C^m(\bar{A})$. Assuming that $A \subset Q$, the Fourier development of $D^p u$ is obtained from the Fourier development of u by formally differentiating it by means of D^p. From this remark it follows that:

8.I. <u>If</u> $u \in H_0^{\sigma}(R)$ <u>and</u> u <u>has every</u> $\mathcal{L}^2$<u>-strong derivative of order</u> m, <u>then</u> $u \in H_m^{\sigma'}(R)$ <u>for any</u> σ' <u>such that</u> $\sigma < \sigma' < \pi$.

In fact, u^* has the $\mathcal{L}^2$-strong derivatives $D^p u^*$ $(|p| = m)$ and if u^* has the following Fourier development in Q,

$$(8.6) \qquad u^* = \frac{1}{(\sqrt{2\pi})^{\tau}} \sum_{-\infty}^{+\infty}{}_{\kappa}\, c_\kappa e^{i\kappa x},$$

then:

$$(8.7) \qquad D^p u^* = \frac{1}{(\sqrt{2\pi})^{\tau}} \sum_{-\infty}^{+\infty}{}_{\kappa}\, i^{|p|} \kappa^p c_\kappa e^{i\kappa x}.$$

It follows that $u^* \in H_m(Q)$. Since u^* vanishes outside of the cylinder $|y| < \sigma$, $|t| < \sigma$, there exists a sequence of C^m functions $\{v_\kappa\}$ having their support in the cylinder $|y| < \sigma'$, $|t| < \sigma'$ and converging toward u^* in the space $H_m(Q)$. Then $u \in H_m^{\sigma'}(R')$.

8.II. <u>If</u> $u \in H_0^{\sigma}(R)$ <u>has the</u> $\mathcal{L}^2$<u>-strong derivative</u> $D^p u = \varphi$ <u>and</u> φ <u>has the</u> $\mathcal{L}^2$<u>-strong derivative</u> $D^q \varphi = \psi$, <u>then</u> u <u>has the</u> $\mathcal{L}^2$<u>-strong derivative</u> $D^q D^p u$ <u>and</u> $D^q D^p = \psi$.

The proof is a trivial consequence of (8.6) and (8.7).

Whenever a function u has some $\mathcal{L}^2$-strong derivative $D^p u$ we simply write $D^p u \in \mathcal{L}^2$.

8.III. Let $u \in H_m^\sigma(R)$. A necessary and sufficient condition for $D_y u \in H_m(R)$ [1] is that for any non-zero real r-vector $h \equiv (h_1, \ldots, h_{r-1}, 0)$ such that $|h| < \pi - \sigma$, the following inequality be satisfied:

$$\left\| \frac{u(x+h) - u(x)}{|h|} \right\|_{m,R}^2 \leq c_1 \tag{8.8}$$

with c_1 independent of h. Then

$$\left\| \frac{u(x+h) - u(x)}{|h|} \right\|_{m,R}^2 = \mathcal{O}\left(\| D_y u \|_{m,R}^2 \right). \tag{8.9}$$

Let us consider the function u^* introduced above. From (8.4), (8.8) it follows that:

$$\left\| \frac{u^*(x+h) - u^*(x)}{|h|} \right\|_{m,Q}^2 \leq c c_1 . \tag{8.10}$$

Let

$$u^*(x) = \frac{1}{(\sqrt{2\pi})^r} \sum_{-\infty}^{+\infty}{}_k \, c_k \, e^{ikx} \tag{8.11}$$

be the Fourier development of u^* in Q. From (8.10) and (8.11)

---

(1) If $u_1, \ldots, u_m$ are the components of the vector u, then by $D_y u$ we mean the $[(r-1) \times m]$-vector $\frac{\partial}{\partial y_h} u_k$ $(h = 1, \ldots, r-1 \; ; \; k = 1, \ldots, m)$.

it follows that

$$(8.12)\quad \sum_{|p|=0}^{m} \sum_{-\infty}^{+\infty}{}_{k}\, k^{2p}|c_k|^2 \left| \frac{e^{ikh}-1}{|h|} \right| \leq \text{constant}.$$

Hence the proof of sufficiency follows by the same argument which was used at the bottom of pag.35.

The proof of (8.9) follows from the inequality :

$$\frac{|e^{ikh}-1|^2}{|h|^2} \leq (k_1^2+\cdots+k_{r-1}^2).$$

Let us now assume that $D_y u \in H_m(R)$. Then $D_y u^* \subset H_m(Q)$. Since u^* vanishes in a strip near ∂Q, any derivative $D^p D_{y_j} u^*$ of u^* ($|p| \leq m$) is given by the development obtained by merely differentiating, with $D^p D_{y_j}$, both sides of (8.11). It follows that the series

$$\sum_{|p|=0}^{m} \sum_{-\infty}^{+\infty}{}_{k}\, k^{2p}|c_k|^2 (k_1^2+\cdots+k_{r-1}^2)$$

is convergent. Hence (8.12) - i.e. (8.10) - holds.

8.IV. Let $u \in H_m^\sigma(R)$ and $D_y u \in H_m(R)$. Set $h = (0,..,h_i,..,0)$, $0 < |h_i| < \sigma$. Then:

$$\lim_{h\to 0} \left\| \frac{u(x+h)-u(x)}{h_i} - \frac{\partial u}{\partial y_i} \right\|_{m,R} = 0.$$

The proof is easily obtained by using the u^* extension of u and Fourier development (8.11).

8.V. Let $\psi \in H_0^{\sigma}(R)$ **and** $w_{\kappa p} \in H_0^{\sigma}(R)$ $(\kappa + |p| \leq \nu)$.
Assume that $D_y \psi \in \mathcal{L}^2$ **and for every** $v \in \overset{\circ}{C}^{\nu+1}(R)$:

$$E(v) \equiv \int_R \psi \frac{\partial^{\nu+1} v}{\partial t^{\nu+1}} dx + \int_R \Big(\sum_{\kappa+|p| \leq \nu} w_{\kappa p} \frac{\partial^{\kappa}}{\partial t^{\kappa}} D_y^p v \Big) dx = 0.$$

Then $\frac{\partial \psi}{\partial t} \in \mathcal{L}^2$, **and** $\| \frac{\partial \psi}{\partial t} \|_{0,R}^2 \leq c_2 \Big(\sum_{\kappa,p} \| w_{\kappa p} \|_{0,R}^2 + \| D_y \psi \|_{0,R}^2 \Big)$.

Let m be any integer not less than $\nu + 1$ and put:

$$\psi^*(y,t) \begin{cases} = \psi(y,t) & \text{for } t > 0, \\ = \sum_{j=1}^{m+1} \lambda_j \psi(y,-jt) & \text{for } t < 0. \end{cases}$$

$$w_{\kappa p}^*(y,t) \begin{cases} = w_{\kappa p}(y,t) & \text{for } t > 0, \\ = \sum_{j=1}^{m+1} \lambda_j (-j)^{\nu+1-\kappa} w_{\kappa p}(y,-jt) & \text{for } t < 0. \end{cases}$$

The λ_j are chosen so that:

$$\sum_{j=1}^{m+1} (-j)^{\nu-\kappa} \lambda_j = 1 \qquad (\kappa = 0, 1, \dots, \nu+1).$$

Let $V \in \overset{\circ}{C}^{\infty}(Q)$. We claim that:

(8.13) $$\tilde{E}(V) \equiv \int_Q \psi^* \frac{\partial^{\nu+1} V}{\partial t^{\nu+1}}\, dx + \int_Q \Big(\sum_{k+|p|\le\nu} w^*_{kp} \frac{\partial^k}{\partial t^k} D^p_y V\Big) dx = 0.$$

In fact we have:

$$\tilde{E}(V) = E(V) + \sum_{j=1}^{m+1} \lambda_j \int_R \psi(y,jt)\, V_{t^{\nu+1}}(y,-t)\, dx +$$

$$+ \sum_{k+|p|\le\nu} \sum_{j=1}^{m+1} \lambda_j (-j)^{\nu+1-k} \int_R w_{kp}(y,jt)\, D^p_y V_{t^k}(y,-t)\, dx =$$

$$= E(V) + \sum_{j=1}^{m+1} \lambda_j j^{-1} \int_R \psi(y,t)\, V_{t^{\nu+1}}(y,-tj^{-1})\, dx +$$

$$+ \sum_{k+|p|\le\nu} \sum_{j=1}^{m+1} \lambda_j j^{-1} (-j)^{\nu+1-k} \int_R w_{kp}(y,t)\, D^p_y V_{t^k}(y,-tj^{-1})\, dx.$$

Set:

$$V_0(y,t) = -\sum_{j=1}^{m+1} \lambda_j (-j)^{\nu} V(y,-tj^{-1}) \ ; \qquad v = V + V_0 .$$

We have:

$$\tilde{E}(V) = E(V+V_0) = E(v)$$

and

$$\left[\frac{\partial^\kappa}{\partial t^\kappa} v(y,t) \right]_{t=0} = 0 \qquad \kappa = 0,1,\ldots,\nu+1.$$

Set:

$$v_s(y,t) \begin{cases} = \eta(t)\, v(y, t-\frac{1}{s}) & \text{for } t \geq \frac{1}{s} \\ = 0 & \text{for } t < \frac{1}{s} \end{cases},$$

where $\eta(t)$ is a C^∞ scalar function which equals 1 in $(0,\sigma)$ and which vanishes identically in a left neighborhood of π. Since $v_s \in \mathring{C}^{\nu+1}(R)$ we have $E v_s = 0$. On the other hand $\lim_{s\to\infty} E v_s = E v$, so that (8.13) is proved.

We now assume that $V = i^{\nu+1}\varphi(x)e^{i\kappa x}\gamma$ where φ is a scalar function belonging to $\mathring{C}^\infty(Q)$ which equals 1 in the cylinder $|y| < \sigma$, $|t| < \sigma$, γ is an arbitrary constant n-vector, and $\kappa \equiv (\ell_1,\ldots,\ell_{n-1},\tau)$ is different from zero. From equation (8.13) we get

$$\tau^{\nu+1}\int_Q \psi^* e^{-i\kappa x}dx = \sum_{0\leq|\alpha|\leq\nu} \int_Q W_\alpha(x)\kappa^\alpha e^{-i\kappa x}dx$$

where the $W_\alpha(x)$ are functions belonging to $\mathcal{L}^2(Q)$.
Let c_κ and $\gamma_{\alpha\kappa}$ be the Fourier coefficients of ψ^* and $W_\alpha(x)$ respectively. We have:

$$|\tau c_\kappa| \le 2^{\nu+1} \frac{|\tau|^{2\nu+3}}{|\kappa|^{2\nu+2}} |c_\kappa| + 2^{\nu+1} \frac{|\tau||\ell|^{2\nu+2}}{|\kappa|^{2\nu+2}} |c_\kappa| \le$$

(8.14)

$$\le \frac{2^{\nu+1}}{|\kappa|^\nu} \sum_\alpha |\kappa|^\nu |\gamma_{\alpha\kappa}| + 2^{\nu+1} |\ell| |c_\kappa| .$$

Since $D_y \Psi^* \in \mathcal{L}^2$, it follows from the above inequality that the series

$$\sum_{-\infty}^{+\infty}{}_\kappa \, \tau^2 |c_\kappa|^2$$

is convergent. Thus $\frac{\partial \Psi^*}{\partial t} \in \mathcal{L}^2$.

From (8.14) it follows that:

$$\left\| \frac{\partial \Psi^*}{\partial t} \right\|_{0,Q}^2 \le c \left(\| W_\alpha \|_{0,Q}^2 + \| D_y \Psi^* \|_{0,Q}^2 \right).$$

Bibliography of Lecture 8

[1] G. FICHERA - See [2] of lecture 1.

[2] K.O. FRIEDRICHS - The identity of weak and strong extensions of differential operators - Trans. Amer. Mathem. Society vol.55, 1944.

[3] J.L. LIONS - Some questions on elliptic equations - Tata Inst. of Fundamental Research, Bombay, 1957.

[4] L. NIRENBERG - On Elliptic Partial Differential Equations - Annali Scuola Norm. Sup. Pisa, 1959.

Lecture 9

Regularity at the boundary: tangential derivatives.

Let A be a bounded domain of X^n such that $\partial A = \partial \bar{A}$. We say that A is C^ν-smooth at the point x^0 of ∂A if a neighborhood I of x^0 exists with the following properties:

i) There exists a C^ν homeomorphism, which maps the set $J = \bar{I} \cap \bar{A}$ onto the closed semiball Σ^+: $t \geq 0$, $|y|^2 + t^2 \leq 1$ of the r-dimensional (y,t) space.

ii) The set $\bar{I} \cap \partial A$ is mapped onto the $(r-1)$-dimensional ball $t = 0$, $|y| \leq 1$.

A is called C^ν-smooth if it is C^ν-smooth at every point of its boundary.

Suppose that u is a solution of equation (7.4) belonging to V (see problem II of lecture 7). Under proper assumptions for V, assuming A is C^ν-smooth in x^0, we shall prove that u has continuous derivatives of any order we wish in $N \cap \bar{A}$, where N is a suitable neighborhood of x^0. Of course ν depends on the order of derivatives we wish to prove exist.

As in the case of the coefficients of L and the function f, we shall suppose, for the sake of brevity, that $\nu = \infty$. We leave it to the reader as an exercise, to determine the more general hypotheses under which the result which we are going to prove, can be obtained with the same method.

Under suitable hypotheses on the operator γ introduced in lecture 7, the results which we shall obtain for equation (7.4) can be easily extended to equation (7.5). However, we shall not discuss this case.

The reader will be able to see, after studying the proof of theor. 9.I of this lecture, what hypotheses are needed on γ in order to extend this proof to the more general case of equation (7.5). On the other hand, the proof of theorem 10.I of the next lecture works exactly the same way in the case of equation (7.5) and requires no additional hypotheses on γ.

It is convenient to collect here all our hypotheses.

1°) <u>The coefficients</u> $a_{pq}(x)$ <u>of the elliptic operator</u> $L \equiv D^q a_{pq} D^q$ <u>and the function</u> $f(x)$ <u>belong to</u> C^∞.

2°) V <u>is a subspace of</u> $H_m(A)$ <u>containing</u> $\overset{\circ}{H}_m(A)$ <u>and such that for every</u> $v \in V$:

(9.1) $$|\mathcal{R}e\, B(v,v)| \geq c_0 \|v\|_m^2 .$$

3°) A <u>is</u> C^∞<u>-smooth at the point</u> x^0 <u>of its boundary</u>.

Let us denote by $\xi \equiv (y,t)$ the point of the (y,t) space and let $\xi = \xi(x)$ be the C^∞-homeomorphism that maps J onto Σ^+ and $x = x(\xi)$ its inverse. For any δ and σ such that $0 < \delta < \sigma < 1$ let $\varphi(\xi)$ be a C^∞ scalar real function, which vanishes outside of the ball Γ_σ: and which equals 1 in the ball $\Gamma_\delta : |\xi| \leq \delta$. Let $v \in V$. Consider the function $w(\xi) = \varphi(\xi) v[x(\xi)]$. We make the following hypothesis on V:

4°) <u>Let</u> $h \equiv (h_1, \ldots, h_{r-1}, 0)$ <u>be a real</u> r<u>-vector such that</u> $0 < |h| < 1-\sigma$. <u>The functions:</u>

$$v_0(x) \begin{cases} = w[\xi(x)] & x \in J \\ = 0 & x \in A-J \end{cases} \qquad v_h(x) \begin{cases} = \dfrac{w[\xi(x)+h]-w[\xi(x)]}{} & x \in J \\ = 0 & x \in A-J \end{cases}$$

belong to V.

Let W be the variety spanned by $v[x(\xi)]$ when v belongs to V and is such that $v = 0$ in $A-J$. It is evident that if $w_0(\xi)$ is a C^∞ function such that $w_0 \in \overset{\circ}{H}_m(\Sigma^+)$ [(1)], then $w_0 \in W$.

5°) $W = \overline{W \cap C^\infty(\Sigma^+)}$ (The closure must be understood in $H_m(\Sigma^+)$).

The hypotheses assumed on V are satisfied if $V \equiv \overset{\circ}{H}_m(A)$ or $V \equiv H_m(A)$.

By considering only functions $v \in V$ such that $v(x) = 0$ in $A-J$, it is possible to write inequality (9.1) in terms of the ξ coordinates. After elementary computations we see that (9.1) can be written as follows in the new coordinates:

(9.2) $$\left| \mathcal{R}e \int_{\Sigma^+} \alpha_{pq}(\xi)\, D^q v\, D^p v\, d\xi \right| \geq C_0 \, \|v\|^2_{m,\Sigma^+}$$

where the $\alpha_{pq}(\xi)$ belong to $C^\infty(\Sigma^+)$ and the symbols D^q, D^q, $\| \ \|$ must be referred to the new coordinates. The function v is now any function of W. From (9.2), by theorem 7.II, it follows that only one solution u exists in the space W of the system

(9.3) $$B(u,v) = (F,v)_0 \qquad v \in W$$

where $F(\xi) = f[x(\xi)] \left| \frac{\partial x}{\partial \xi} \right|$. We have maintained the symbol B to denote

---

(1) According to our definition of the spaces $\overset{\circ}{H}_m$ we should write $\overset{\circ}{H}_m(\Sigma^+ - \partial\Sigma^+)$ instead of $\overset{\circ}{H}_m(\Sigma^+)$. But this would be quite a pedantry.

the new bilinear form. The function $F(\xi)$ belongs to $C^\infty(\Sigma^+)$.

9.I. <u>Let</u> δ <u>be any positive number such that</u> $0<\delta<1$. <u>Let</u> Γ_δ^+ <u>be the semiball:</u> $t\geq 0$, $|y|^2+t^2\leq\delta^2$. <u>Let</u> $s\equiv(s_1,\dots,s_{r-1})$ <u>be an arbitrary</u> $(r-1)$<u>-vector-index. The solution</u> u <u>of (9.3) is such that</u> $D_y^s u \in H_m(\Gamma_\delta^+)$. <u>Moreover</u>

$$\text{(9.4)}\qquad \|D_y^s u\|^2_{m,\Gamma_\delta^+} \leq c_1\left(\|F\|^2_{|s|-m,\Sigma^+}+\|u\|^2_{m,\Sigma^+}\right)$$

<u>with</u> c_1 - <u>for any given</u> s - <u>only depending on the</u> α_{pq} <u>'s and</u> δ.

The theorem is true for $|s|=0$. We shall prove the theorem for $|s|=k+1$ supposing it to be true for $|s|\leq k$.

Let $\varphi(x)$ [2] be the C^∞ scalar function introduced above. Because of hypothesis 4°) and the induction hypothesis, the function $D_y^k\varphi u$ belongs to W (see lemma 8.IV); D_y^k <u>is any</u> y<u>-partial differentiation of order</u> k.

Set $U(x)=D_y^k\varphi u$. We have for $v\in W\cap C^\infty(\Sigma^+)$ and any real $h\equiv(h_1,\dots,h_{r-1},0)$ such that $0<|h|<\frac{1-\delta}{2}$:

$$B\left(\frac{U(x+h)-U(x)}{|h|},v\right)=(-1)^k\int_{\Sigma^+}D^q\,\frac{\varphi(x+h)u(x+h)-\varphi(x)u(x)}{|h|}\,.$$

$$\cdot D_y^k(\bar\alpha_{pq}D^p v)dx=(-1)^k\int_{\Sigma^+}\alpha_{pq}D^q\,\frac{\varphi(x+h)u(x+h)-\varphi(x)u(x)}{|h|}\,D_y^k D^p v\,dx+$$

---

(2) From now on we shall use the letter x in place of ξ.

$$+ \sum_{0\le|j|\le K-1} \int_{\Sigma^+} D_y^j \left[\beta_{pq}^j D^q \frac{\varphi(x+h)u(x+h)-\varphi(x)u(x)}{|h|}\right] D^p v\, dx .$$

The meaning of the symbol β_{pq}^j is self-explanatory.

Due to the induction hypothesis and lemma 8.III, the last integral is $O\left(\sum_{|s|\le K} \|D_y^s u\|_{m,\Gamma^+_{\sigma'}} \|v\|_m\right)$ (3), with $\sigma' = \frac{1+\sigma}{2}$.

We have:

$$(-1)^K \int_{\Sigma^+} \alpha_{pq} D^q \frac{\varphi(x+h)u(x+h)-\varphi(x)u(x)}{|h|} D_y^K D^p v\, dx =$$

$$= (-1)^K \int_{\Sigma^+} \alpha_{pq} \frac{\varphi(x+h)D^q u(x+h)-\varphi(x)D^q u(x)}{|h|} D^p D_y^K v\, dx +$$

$$+ \int_{\Sigma^+} D_y^K \left[\alpha_{pq} \sum_{\substack{q',q'' \\ |q''|<m}} \frac{D^{q'}\varphi(x+h)D^{q''}u(x+h)-D^{q'}\varphi(x)D^{q''}u(x)}{|h|}\right] D^p v\, dx =$$

$$= (-1)^K \int_{\Sigma^+} \frac{\alpha_{pq}(x+h)\varphi(x+h)D^q u(x+h)-\alpha_{pq}(x)\varphi(x)D^q u(x)}{|h|} D^p D_y^K v\, dx$$

---

(3) By simply writing $\| \quad \|_m$ we mean the norm over Σ^+.

$$- \int_{\Sigma^+} D_y^k \left[\frac{a_{pq}(x+h) - a_{pq}(x)}{|h|} \varphi(x+h) D^q u(x+h) \right] D^p v \, dx +$$

$$+ O\left(\sum_{|s| \leq k} \| D_y^s u \|_{m, \Gamma_{\sigma'}^+} \| v \|_m \right) =$$

$$= (-1)^k \int_{\Sigma^+} a_{pq}(x) D^q u(x) \varphi(x) D^p D_y^k \frac{v(x-h) - v(x)}{|h|} dx +$$

$$+ O\left(\sum_{|s| \leq k} \| D_y^s u \|_{m, \Gamma_{\sigma'}^+} \| v \|_m \right) =$$

$$= (-1)^k \int_{\Sigma^+} a_{pq}(x) D^q u(x) D^p \left[\varphi(x) D_y^k \frac{v(x-h) - v(x)}{|h|} \right] dx +$$

$$+ O\left(\sum_{|s| \leq k} \| D_y^s u \|_{m, \Gamma_{\sigma'}^+} \| v \|_m \right).$$

Then we have:

$$B\left(\frac{U(x+h) - U(x)}{|h|}, v \right) = (-1)^k B\left(u, \varphi(x) D_y^k \frac{v(x-h) - v(x)}{|h|} \right) +$$

$$+ O\left(\sum_{|s| \leq k} \| D_y^s u \|_{m, \Gamma_{\sigma'}^+} \| v \|_m \right).$$

It is

$$B\left(u, \varphi(x) D_y^{\kappa} \frac{v(x-h)-v(x)}{|h|}\right) = \int_{\Sigma^+} \varphi F D_y^{\kappa} \frac{v(x-h)-v(x)}{|h|} dx =$$

$$= O\left(\|F\|_{\kappa-m+1} \left\| \frac{v(x-h)-v(x)}{|h|} \right\|_{m-1}\right) = O\left(\|F\|_{\kappa-m+1} \|v\|_m\right).$$

For any $v \in W$, we get

$$B\left(\frac{U(x+h)-U(x)}{|h|}, v\right) = O\left[\left(\sum_{|s|\leq \kappa} \|D_y^s u\|_{m, \Gamma_{\sigma'}^+} + \|F\|_{\kappa-m+1}\right) \|v\|_m\right].$$

By assuming $v = \dfrac{U(x+h)-U(x)}{|h|}$, we obtain from (9.2):

$$\left\| \frac{U(x+h)-U(x)}{|h|} \right\|_m \leq c_2 \left(\sum_{|s|\leq \kappa} \|D_y^s u\|_{m, \Gamma_{\sigma'}^+} + \|F\|_{\kappa-m+1}\right).$$

From this estimate the proof follows.

Bibliography of Lecture 9

[1] F.E. BROWDER - see [1],[2] of lecture 5.

[2] F.E. BROWDER - A priori estimates for solutions of elliptic boundary value problems, I & II - Indag. Math. vol.22, 1960.

[3] F.E. BROWDER - Estimates and existence theorems for elliptic boundary value problems - Proc.Nat.Acad.Sci. U.S.A., vol.45, 1959.

[4] G. FICHERA - Linear elliptic equations of higher order in two independent variables and singular integral equations, with applications to anisotropic inhomogeneous elasticity - Proceedings of the Int. Conference on PDE and Continuum Mechanics, Madison, Wis. 1960 - R.E.Langer Editor.

[5] L. HÖRMANDER - On the regularity of the solutions of boundary problems - Acta Mathem. vol. 99, 1958.

[6] L. NIRENBERG - see [4] of lecture 4 and [4] of lecture 8.

[7] A.J. VOLPERT - Boundary value problems for elliptic systems of differential equations of higher order in the plane - Dokl. Akad. Nauk SSSR v.127, 1959 (in Russian).

Lecture 10

Regularity at the boundary : final results.

In lecture 9 we proved that the solution u of (9.3) has any $\mathcal{L}^2$-strong derivative $D_y^s D^p u$ in the semiball Γ_δ^+ for any s and any $|p| = m$. We wish now to complete this result and prove that u has any $\mathcal{L}^2$-strong derivative of arbitrary order in Γ_δ^+. This result is given by the following theorem.

10.I. Given an arbitrary positive $\delta < 1$, the solution u of (9.3) has, for any s and any $i > 0$, the $\mathcal{L}^2$-strong derivative $\frac{\partial^{m+i}}{\partial t^{m+i}} D_y^s u$ in Γ_δ^+. A constant c_2 exists depending only on the α_{pq}'s and on δ such that:

$$\left\| \frac{\partial^{m+i}}{\partial t^{m+i}} D_y^s u \right\|_{0,\Gamma_\delta^+} \leq c_2 \left(\|F\|^2_{|s|+i-m,\Sigma^+} + \|u\|^2_{m,\Sigma^+} \right). \tag{10.1}$$

Let us suppose that:

i_1) $\frac{\partial^{m+j}}{\partial t^{m+j}} D_y^s u \in \mathcal{L}^2(\Gamma_\delta^+)$ $\quad (j = 0, \dots, i-1)$

for any s and any $\delta < 1$.

This is true for $i=1$. Set $U(x)=\varphi(x)\,u(x)$, where $\varphi(x)$ is the scalar function used in the previous lecture. For $v \in \overset{\circ}{C}{}^{\infty}(R)$ we have, with an obvious meaning for β^{κ}_{pq}:

$$B\left(U, \frac{\partial^{i-1}}{\partial t^{i-1}} v\right) = (-1)^{i-1} \int_{\Sigma^+} \alpha_{pq} \frac{\partial^{i-1}}{\partial t^{i-1}} D^q U\, D^p v\, dx +$$

$$+ \sum_{\kappa=0}^{i-2} \int_{\Sigma^+} \beta^{\kappa}_{pq} \frac{\partial^{\kappa}}{\partial t^{\kappa}} D^q U\, D^p v\, dx = (-1)^{i-1} \int_{\Sigma^+} \alpha_{pq} \frac{\partial^{i-1}}{\partial t^{i-1}} D^q U\, D^p v\, dx +$$

$$+ O\left(\sum_{j+|l| \le m+i-1} \left\| \frac{\partial^j}{\partial t^j} D^l_y u \right\|_{0,\Gamma^+_\sigma} \|v\|_{m-1} \right).$$

Let $\tilde{p} \equiv (0,\dots,0,m)$. We denote by $\sum\limits_{pq}^{(\tilde{p})}$ a summation extended to any pair of vector-indeces p,q $(0 \le |p| \le m,\ 0 \le |q| \le m)$ excluding the pair $\tilde{p},\tilde{p}$.

We have, using symbols whose meaning is self-explanatory:

$$B\left(U, \frac{\partial^{i-1}}{\partial t^{i-1}} v\right) = \int_{\Sigma^+} \alpha_{pq} D^q u\, D^p\left(\varphi \frac{\partial^{i-1}}{\partial t^{i-1}} v\right) dx +$$

$$+ \sum_{\substack{|p| \le m \\ |\kappa| < m}} \int_{\Sigma^+} \beta_{p\kappa} D^{\kappa} u\, D^p \frac{\partial^{i-1}}{\partial t^{i-1}} v\, dx +$$

$$+ \sum_{\substack{|k|<m \\ |q|\leq m}} \int_{\Sigma^+} \gamma_{kq} D^q u \, D^k \frac{\partial^{i-1}}{\partial t^{i-1}} v \, dx =$$

$$= \left(F, \varphi \frac{\partial^{i-1}}{\partial t^{i-1}} v\right)_0 + O\left(\sum_{j+|\ell|\leq m+i-1} \left\| \frac{\partial^j}{\partial t^j} D_y^\ell u \right\|_{0,\Gamma_\sigma^+} \|v\|_{m-1} \right);$$

$$\int_{\Sigma^+} \alpha_{pq} \frac{\partial^{i-1}}{\partial t^{i-1}} D^q U \, D^p v \, dx = \int_{\Sigma^+} \alpha_{\tilde{p}\tilde{p}} \frac{\partial^{m+i-1}}{\partial t^{m+i-1}} U \frac{\partial^m v}{\partial t^m} dx +$$

$$+ \sum_{p,q}^{(\tilde{p})} \int_{\Sigma^+} \alpha_{pq} \frac{\partial^{i-1}}{\partial t^{i-1}} D^q U \, D^p v \, dx =$$

$$= \int_{\Sigma^+} \alpha_{\tilde{p}\tilde{p}} \frac{\partial^{m+i-1}}{\partial t^{m+i-1}} U \frac{\partial^m v}{\partial t^m} dx + O\left(\sum_{\substack{j+|\ell|\leq m+i \\ j\leq m+i-1}} \left\| \frac{\partial^j}{\partial t^j} D_y^\ell u \right\|_{0,\Gamma_\sigma^+} \|v\|_{m-1} \right).$$

We deduce that:

$$\int_{\Sigma^+} \alpha_{\tilde{p}\tilde{p}} \frac{\partial^{m+i-1}}{\partial t^{m+i-1}} U \frac{\partial^m v}{\partial t^m} dx =$$

$$= O\left(\left\{ \sum_{\substack{j+|\ell|\leq m+i \\ j\leq m+i-1}} \left\| \frac{\partial^j}{\partial t^j} D^\ell u \right\|_{0,\Gamma_\sigma^+} + \|F\|_{i-m} \right\} \|v\|_{m-1} \right).$$

From lemma 8.V and since $\det \alpha_{\tilde{p}\tilde{p}} \neq 0$ (the operator is elliptic!) it follows $\frac{\partial^{m+i}}{\partial t^{m+i}} U \in \mathcal{L}^2$ and moreover that:

$$(10.2) \quad \left\| \frac{\partial^{m+i}}{\partial t^{m+i}} u \right\|^2_{0,\Gamma^+_\delta} \leq c_4 \left\{ \sum_{\substack{j+|\ell| \leq m+i \\ j \leq m+i-1}} \left\| \frac{\partial^j}{\partial t^j} D^\ell u \right\|^2_{0,\Gamma^+_\sigma} + \| F \|^2_{i-m} \right\} .$$

Let us suppose that we have, together with i_1), the following induction hypothesis:

i_2)
$$\frac{\partial^{m+i}}{\partial t^{m+i}} D^s_y u \in \mathcal{L}^2(\Gamma'^+_\delta)$$
<u>for</u> $|s| \leq k$ <u>and any</u> $\delta < 1$.

We have proven that i_2) is true for $k = 0$. We shall prove that:

$$\frac{\partial^{m+i}}{\partial t^{m+i}} D^s_y u \in \mathcal{L}^2(\Gamma^+_\delta)$$

for $|s| = k+1$.

We have for any D^s_y such that $|s| = k+1$

$$B\left(U, \frac{\partial^{i-1}}{\partial t^{i-1}} D^s_y v\right) = (-1)^{k+i} \int_{\Sigma^+} \alpha_{pq} \frac{\partial^{i-1}}{\partial t^{i-1}} D^s_y D^q U D^p v \, dx +$$

$$+ \sum_{\substack{j+|h| < k+i \\ j \leq i-1}} \int_{\Sigma^+} \beta^{jh}_{pq} \frac{\partial^j}{\partial t^j} D^h_y D^q U D^p v \, dx =$$

$$= (-1)^{k+i} \int_{\Sigma^+} \alpha_{pq} \frac{\partial^{i-1}}{\partial t^{i-1}} D_y^s D^q U D^p v \, dx + O\Big(\sum_{\substack{j+|\ell| \le m+k+i \\ j \le m+i}} \Big\| \frac{\partial^j}{\partial t^j} D_y^\ell u \Big\|_{0,\Gamma_\sigma^+} \| v \|_{m-1} \Big).$$

By using the same arguments as before ,

$$B\Big(U, \frac{\partial^{i-1}}{\partial t^{i-1}} D_y^s v\Big) = \Big(F, \varphi \frac{\partial^{i-1}}{\partial t^{i-1}} D_y^s v\Big)_0 +$$

$$+ O\Big(\sum_{\substack{j+|\ell| \le m+k+i \\ j \le m+i-1}} \Big\| \frac{\partial^j}{\partial t^j} D_y^\ell u \Big\|_{0,\Gamma_\sigma^+} \| v \|_{m-1} \Big) ;$$

$$\int_{\Sigma^+} \alpha_{pq} \frac{\partial^{i-1}}{\partial t^{i-1}} D_y^s D^q U D^p v \, dx = \int_{\Sigma^+} \alpha_{\tilde p \tilde p} \frac{\partial^{m+i-1}}{\partial t^{m+i-1}} D_y^s U \frac{\partial^m v}{\partial t^m} dx +$$

$$+ O\Big(\sum_{\substack{j+|\ell| \le m+k+1+i \\ j \le m+i-1}} \Big\| \frac{\partial^j}{\partial t^j} D^\ell u \Big\|_{0,\Gamma_\sigma^+} \| v \|_{m-1} \Big).$$

Then we deduce that the $\mathcal{L}^2$**-strong derivative** $\dfrac{\partial^{m+i}}{\partial t^{m+i}} D_y^s u$ **exists and moreover that:**

$$(10.3)\qquad \left\| \frac{\partial^{m+i}}{\partial t^{m+i}} D_y^{s} u \right\|^2_{0,\Gamma^+_\delta} \le c_5 \left\{ \sum_{\substack{j+|l|\le m+k+i \\ j\le m+i}} \left\| \frac{\partial^j}{\partial t^j} D_y^{l} u \right\|^2_{0,\Gamma^+_\sigma} + \sum_{\substack{j+|l|\le m+k+1+i \\ j\le m+i-1}} \left\| \frac{\partial^j}{\partial t^j} D_y^{l} u \right\|^2_{0,\Gamma^+_\sigma} + \| F \|^2_{|s|+i-m} \right\}.$$

Suppose we have already proven that (10.1) holds for the derivatives considered in the induction hypothesis i_1). Then, by using (10.2) we prove, by induction, that (10.1) is true for the derivatives considered in the induction hypothesis i_2). From (10.3) it follows that (10.1) is true in any case.

10.II. **<u>For any</u> $\delta < 1$ <u>the solution</u> u <u>of (9.3) belongs to</u> $H_l(\Gamma^+_\delta)$ <u>with</u> l <u>given arbitrarily and</u>**

$$\| u \|^2_{l,\Gamma^+_\delta} \le c \, \| F \|^2_{l-2m,\Sigma^+} .$$

The constant c depends only on α_{pq}'s , δ , l , and on the constant C_0 of (9.2).
The proof is a trivial consequence of theorems 9.I, 10.I and of (9.2) and (9.3).

Let us now consider again the domain A and assume that it is C^∞-smooth at any point of its boundary and V satisfies conditions of lecture 9 for any $x^\circ \in \partial A$. Suppose that inequality (9.1) holds. Then as a consequence of theorem 9.I, theorem 10.II and of (9.1), it follows that:

10.III. <u>For any</u> $u \in C^\infty(\bar{A})$ <u>and for any given</u> ℓ <u>the following inequality holds:</u>

(10.4) $$\|u\|_\ell^2 \le c\,\|Lu\|_{\ell-2m}^2 ,$$

<u>the constant</u> c <u>depends only on</u> A, ℓ, C_0, <u>and on the</u> a_{pq}<u>'s; it may be explicitly calculated.</u>

Concerning the last statement of this theorem, the reader is requested to notice that in the proof of any inequality we have considered, for the constants $c_1, c_2, \ldots$ explicit expressions can be given. Hence an explicit expression could be given for c of (10.4). This would be extremely tedious but completely feasible.

By the Sobolev lemma, the solution of problem II (see lecture 7) belongs to $C^\infty(\bar{A})$, provided $f \in C^\infty(\bar{A})$.

For any $u \in C^\infty(\bar{A})$ and any $v \in H_m(A)$ the following Green's formula holds:

$$B(u,v) = \int_A (Lu)\,v\,dx + \int_{\partial A} H(u,v)\,d\sigma$$

where $H(u,v)$ is a bilinear form in u and v containing derivatives of v of order not exceeding $m-1$. We define the linear manifold $M(V)$ as follows. A function u belongs to $M(V)$ if and only if: i) $u \in V \cap C^\infty(\bar{A})$; ii) $\int_{\partial A} H(u,v)\,d\sigma = 0$ for any $v \in V$.

10.IV. <u>In the assumed hypotheses for</u> A, L, f <u>and</u> V, <u>the function</u> u <u>is a solution of problem II of lecture 7, if and only if</u>

it is a solution of the following problem:

$$Lu = f, \qquad u \in M(V).$$

The proof is obvious.

In the case where $V = \mathring{H}_m(A)$ the variety $M(V)$ is composed of all the functions which belong to $C^\infty(\bar{A})$ and are such that $D^p u = 0$ on ∂A ($|p| \leq m-1$).

Let us denote by Gf the solution of problem II. G is called the Green's transformation of the problem. For any $\ell \geq 2m$, G is a bounded linear transformation from $H_{\ell-2m}(A)$ into $H_\ell(A)$. Assuming $\ell = 2m$, we see that we may regard G as a compact transformation of H_0 into itself (see theor. 3.IV). Moreover G maps any C^∞ function into a C^∞ function. If L is formally self-adjoint [1], i.e. $L^* = L$, then, as is easily seen, G is hermitian, i.e.

$$(Gf, h)_0 = (h, Gh)_0 .$$

Let us now consider the following problem:

$$Lu + \lambda u = f, \tag{10.5}$$

where λ is any complex constant. The hypotheses on A, L, f, V are the above specified ones. The following problem is perfectly equivalent to problem (10.5):

$$\varphi + \lambda G\varphi = f \qquad \qquad \varphi \in C^\infty(\bar{A}) \tag{10.6}$$

---

[1] When we say that L is formally self-adjoint we mean $a_{pq}(x) \equiv (-1)^{|p|+|q|} \bar{a}_{qp}(x)$

in the sense that if φ is a solution of (10.6) then $u = G\varphi$ is a solution of (10.5) and conversely, to any solution u of (10.5) there corresponds a solution $\varphi = Lu$ of (10.6).

If $f \in C^\infty(\bar{A})$, then any $\mathcal{L}^2$ solution φ of the equation (10.6) necessarily belongs to $C^\infty(\bar{A})$. In fact, if $\varphi \in \mathcal{L}^2(A)$, then $\varphi = f - \lambda G \varphi$ belongs to $H_{2m}(A)$ and hence to $H_{4m}(A)$... ... and so on. Thus we can consider equation (10.6) in the space $\mathcal{L}^2(A)$ where we can apply the Riesz-Fredholm theory. We are thereby led to the following theorem:

10.V. We have one and only one solution of problem (10.5) whenever λ is not an eigenvalue for this problem, i.e. whenever λ is such that, assuming $f \equiv 0$, (10.5) has only the trivial solution $u \equiv 0$. The set of these eigenvalues is a set with no limit-points in any bounded set of the complex plane.

We wish now to show, as a further application, how it is possible to give an existence and uniqueness theorem for linear integro-differential equations connected with elliptic operators. These equations were introduced by Volterra more than fifty years ago, in the foundation of his theory of hereditary elasticity, and have been reconsidered also recently by some authors in connection, for instance, with linear visco-elasticity.

Let us assume that A is C^∞-smooth and that the above mentioned hypotheses on the coefficients $a_{pq}(x)$ of the elliptic operator L and on V hold.

For any $x \in A$ and any t in the closed interval $(0,T)$ let us consider the integro-differential equation

$$(10.7) \quad Lu(x,t) + \sum_{|\alpha| \leq 2m} \int_0^t K_\alpha(t,\tau) D^\alpha u(x,\tau)\, d\tau = f(x,t).$$

The derivatives D^α must be understood to be taken with respect to the space-like coordinates $x_1, \ldots, x_r$. The matrices $K_\alpha(t,\tau)$ are complex $n \times n$ matrices, continuous over the closed square $(0,T) \times (0,T)$. The function $f(x,t)$ is supposed to be C^∞ as a function of x for any $t \in (0,T)$ and continuous together with all of its space-like derivatives in $\bar{A} \times (0,T)$. Of course, what we are going to say holds under more general hypotheses for A, L and f.

The problem consists in proving that there exists one and only one solution $u(x,t)$ of (10.7) which, for any given $l \geq 0$, when regarded as a function of t with values in the space $H_l(A)$, is a continuous function of t whose values belong to $M(V)$.

Set $Lu(x,t) = \varphi(x,t)$. It is readily seen that the following integral equation in the new unknown function $\varphi(x,t)$

(10.8) $$\varphi(x,t) + \sum_{|\alpha| \leq 2m} \int_0^t K_\alpha(t,\tau) D^\alpha G\varphi(x,\tau)\, d\tau = f(x,t)$$

is equivalent to (10.7) with the "boundary condition" $u(x,t) \in M(V)$ for any $t \in (0,T)$. This means that to any $\varphi(x,t)$ which is a solution of (10.8), there corresponds a solution $u(x,t)$ of (10.7) given by $G\varphi(x,t)$ and belonging - as a function of x - to $M(V)$. Conversely, to such a $u(x,t)$ there corresponds a solution $\varphi(x,t) = Lu(x,t)$ of (10.8) belonging to H_{l-2m} for any $l \geq 2m$.

Let us write (10.8) as follows:

(10.9) $$\Phi(t) + \int_0^t N(t,\tau)\, \Phi(\tau)\, d\tau = F(t),$$

where $\Phi(t)$ and $F(t)$ must be understood to be functions of t with values in $H_k(A)$ ($k = \ell - 2m$), $N(t,\tau)$ is, for any $(t,\tau) \in$ $\in (0,T) \times (0,T)$, a linear bounded operator from $H_k(A)$ in itself, such that $\| N(t,\tau) \| \leq C$ (C constant).

The classical successive approximations method, used for the "scalar" Volterra integral equation works in exactly the same way for equation (10.9) and proves the existence of one and only one solution for (10.8), i.e. for our integro-differential problem.

Bibliography of Lecture 10

See bibliography of lecture 9.

Lecture 11

The classical elliptic BVP of Mathematical Physics: 2nd order linear PDE.

Let us consider a 2nd order linear elliptic equation with real coefficients:

$$(11.1)\qquad Lu \equiv \frac{\partial}{\partial x_i}\left[a_{ij}(x)\frac{\partial u}{\partial x_j}\right] + b_i(x)\frac{\partial u}{\partial x_i} + c(x)u = f(x).$$

The unknown function u is now a real-valued function. The functions a_{ij}, b_i, c, f are supposed to belong to C^∞. The bounded domain A is supposed to be C^∞-smooth. These hypotheses will be maintained throughout this lecture. We assume that L is elliptic and positive in $\bar{A}$. This means that the real quadratic form $a_{ij}(x)\xi_i\xi_j$ is positive definite for every $x \in \bar{A}$.

Let us first consider the Dirichlet problem for (11.1). As a consequence of the theory developed in the previous lectures, we have the following existence and uniqueness theorem for this problem.

11.I. Under the above-mentioned hypotheses on L and A and under the further assumption that $c(x) \leq 0$ for any $x \in A$, there exists one and only one C^∞ function $u(x)$ such that:

(11.2) $Lu = f$ in A , (11.3) $u = 0$ on ∂A.

Let us first suppose that for $x \in A$,

(11.4) $$\frac{1}{2}\frac{\partial b_i}{\partial x_i} - c \geq 0.$$

We have for any $v \in \overset{\circ}{H}_1(A)$:

$$-B(v,v) = \int_A \left[a_{ij}\frac{\partial v}{\partial x_i}\frac{\partial v}{\partial x_j} + \left(\frac{1}{2}\frac{\partial b_i}{\partial x_i} - c\right)v^2\right]dx \geq c_0 \|v\|_1^2,$$

as follows from the ellipticity of L and from lemma 3.I. Thus the theorem is proven under the assumption (11.4). The proof in the case $c(x) \leq 0$ follows from theor. 10.V and from the inequality $|u(x)| \leq \max_{x \in \partial A} |u(x)|$ which – as is well known – holds for any smooth solution of the homogeneous equation $Lu = 0$ when $c(x) \leq 0$ (see, for instance, [7], p. 4–5).

Let us now consider another classical BVP for the elliptic equation (11.1), the so called <u>regular oblique derivative problem.</u> Let $\lambda \equiv (\lambda_1, \dots, \lambda_r)$ be a real unit vector defined for any $x \in \partial A$ which is C^∞ when considered as a function of the point x varying on ∂A. The following BVP is known as the oblique derivative problem for the operator L :

(11.5) $Lu = f$ in A , (11.6) $\dfrac{\partial u}{\partial \lambda} = 0$ on ∂A.

Let $\nu \equiv (\nu_1, \dots, \nu_r)$ be the interior unit normal to ∂A . Under the further assumption $\lambda\nu > 0$, the problem (11.5) (11.6) is said to be <u>regular</u>.

11.II. <u>If</u> $c(x) < 0$ <u>for</u> $x \in \bar{A}$, <u>then there exists one and only one</u> C^∞ <u>solution of the regular oblique derivative problem (11.5) (11.6)</u>.

Let ϱ_{ij} be arbitrary functions belonging to $C^\infty(\bar{A})$ such that $\varrho_{ij} = -\varrho_{ji}$. The operator L can be written as follows:

$$(11.1') \qquad Lu = \frac{\partial}{\partial x_i}\left[\alpha_{ij}\frac{\partial u}{\partial x_j}\right] + \beta_i \frac{\partial u}{\partial x_i} + cu ,$$

where:

$$\alpha_{ij} = a_{ij} + \varrho_{ij} \quad , \quad \beta_i = b_i - \frac{\partial \varrho_{ji}}{\partial x_j}$$

It is possible to choose the functions ϱ_{ij} in such a way that:

$$(11.7) \qquad \alpha_{ij}\nu_i = \varrho\lambda_j \qquad (j = 1,\dots,r)$$

where ϱ is a positive C^∞ scalar function defined on ∂A. In order to prove this, let us fix arbitrarily the point x on ∂A. Let $\nu^1,\dots,\nu^r$ be an orthonormal system of vectors such that ν^1 coincides with the interior normal vector ν at ∂A in x. We may suppose that the functions $\nu^1,\dots,\nu^r$, λ are defined throughout $\bar{A}$ and belong to $C^\infty(\bar{A})$. There is no loss in generality in assuming this, since we can always choose $\nu^2,\dots,\nu^r$ belonging to $C^\infty(\partial A)$ and thereafter continue all the above $r+1$ functions throughout $\bar{A}$. We can also suppose that $\lambda\nu^1$ is positive in $\bar{A}$ since, by the regularity condition, it is positive on ∂A. Let us consider the functions R_{hk} defined as follows in $\bar{A}$:

$$R_{hk} \begin{cases} = a_{ij}\nu_i^1\nu_j^1 \dfrac{\lambda\nu^k}{\lambda\nu^1} - a_{ij}\nu_i^1\nu_j^k & (\text{for } h=1 \text{ and } h\le k) \\ = 0 & (\text{for } 1<h\le k) \\ = -R_{kh} & (\text{for } h>k). \end{cases}$$

The functions:

$$\varrho_{ij} = R_{hk}\nu_i^h\nu_j^k$$

satisfy equations (11.7) with ϱ given by:

$$\varrho = \frac{a_{ij}\nu_i^1\nu_j^1}{\lambda\nu^1}.$$

Let us consider for $u, v \in H_1(A)$ the bilinear form:

$$B(u,v) = \int_A \left\{ -\alpha_{ij}\frac{\partial v}{\partial x_i}\frac{\partial u}{\partial x_j} + \beta_i v \frac{\partial u}{\partial x_i} + c v u \right\} dx.$$

Let p_0, p_1, p_2 be positive numbers such that for every $x \in \bar{A}$ we have:

$$\alpha_{ij}(x)\lambda_i\lambda_j > p_0|\lambda|^2, \quad |\beta_i| \le p_1 \quad (i=1,2,\dots,r), \quad -c \ge p_2$$

It follows that:

$$-B(v,v) \geq \int_A \left(p_0 \sum_{i=1}^{n} \left|\frac{\partial v}{\partial x_i}\right|^2 - p_1 |v| \sum_{i=1}^{n} \left|\frac{\partial v}{\partial x_i}\right| + p_2 |v|^2\right) dx.$$

Then for p_2 large enough,

$$-B(v,v) \geq c_0 \|v\|_1^2$$

for any $v \in H_1(A)$. From this inequality follows the existence of a unique function $u \in H_1(A)$ satisfying the equation:

$$B(u,v) = \int_A f v \, dx \tag{11.8}$$

for any $v \in H_1(A) \equiv V$. Since $u \in C^\infty(\bar{A})$ we have, by (11.7),

$$B(u,v) = \int_A v L u \, dx + \int_{\partial A} v \rho \frac{\partial u}{\partial \lambda} d\sigma. \tag{11.9}$$

Due to the arbitrariness of v, it follows from (11.8), (11.9) that u is a solution of (11.5),(11.6). In other words, the class $M(V)$ is, in the present case, composed of the C^∞ functions which satisfy the boundary condition (11.6). In order to prove our theorem, we need only show that the solution of the regular oblique derivative BVP is unique if $c < 0$ on $\bar{A}$. In fact, the existence will follow by using theor.10.V. The desired uniqueness is a consequence of well-known results on linear elliptic PDE's of second order.

Since $c < 0$, it follows that for any smooth non vanishing solution of $Lu = 0$ in the (connected) domain A : $|u(x)| < \max_{x \in \partial A} |u|$

for any $x \in A$. On the other hand, if x^0 is a maximum (minimum) for such a solution u on ∂A, by a theorem of Giraud, we must have $\left[\frac{\partial u}{\partial \lambda}\right]_{x=x^0} < 0$ $\left\{\left[\frac{\partial u}{\partial \lambda}\right]_{x=x^0} > 0\right\}$ (see [7] p. 4-5). From this the above mentioned uniqueness follows.

If we choose the ρ_{ij} in such a way that $\alpha_{ij} = \alpha_{ji}$, then the oblique direction λ given by (11.7) is known as the conormal direction and the resulting BVP is called the Neumann BVP.

Let us now suppose that $\partial_1 A$ and $\partial_2 A$ are two disjoint subsets of ∂A such that, if x^0 is any point of $\partial_k A$ $(k = 1,2)$, the neighborhood I of x^0 enjoying the properties stated at the beginning of lecture 9, can be chosen in such a way that $\bar{I} \cap \partial A \subset \partial_k A$. Moreover, we assume that $\overline{\partial_1 A} = \partial A - \partial_2 A$. Let λ be the C^∞ unit vector defined on ∂A (already introduced in the oblique derivative problem) which satisfies the regularity condition $\lambda \nu > 0$ on ∂A.

The following BVP is known as the mixed BVP:

(11.10) $$Lu = f \quad \text{in } A, \qquad u = 0 \text{ on } \overline{\partial_1 A}, \qquad \frac{\partial u}{\partial \lambda} = 0 \quad \text{on } \partial_2 A.$$

As space V we take now the closure in the space $H_1(A)$ of the subclass of $C^1(\bar{A})$ composed of the functions which vanish on $\overline{\partial_1 (A)}$.

We use the representation (11.1') of Lu with the coefficients α_{ij} satisfying (11.7). By the same argument used in the oblique derivative problem, we see that:

$$-B(v,v) \geq c_0 \|v\|_1^2$$

for any $v \in V$, when $-c \geq p_2$ with p_2 large enough. Then we have a unique solution of (11.8), corresponding to the present choice of V.

The class V satisfies the conditions for the regularisation theory of lectures 9 and 10 in the neighborhood of any point of $\partial_1 A$ and of $\partial_2 A$. Hence, the solution of the mixed BVP has the following regularity properties:

i) u belongs to $C^{\infty}(A) \cap H_1(A)$.

ii) u belongs to $C^{\infty}(A \cup \partial_1 A \cup \partial_2 A)$.

If $\partial A = \partial_1 A \cup \partial_2 A$, then u is C^{∞} in the closed domain $\bar{A}$, otherwise the only points where u could not be C^{∞} are in the set $\partial A - (\partial_1 A \cup \partial_2 A)$.

It has been proven that u is continuous in the closed domain $\bar{A}$ (see [3]).

Bibliography of Lecture 11.

[1] G.BOULIGAND-G.GIRAUD-P.DELENS - Le probléme de la dérivée oblique en théorie du potentiel - Actual. Scient. Industr. Hermann, Paris, 1935.

[2] G.FICHERA - Sul problema della derivata obliqua e sul problema misto per l'equazione di Laplace - Bollett. Unione Mat. Ital. 1952.

[3] G.FICHERA - Alcuni recenti sviluppi della teoria dei problemi al contorno, etc - Atti Conv. Intern. sulle Eq. Der. Parz. 1954 - Cremonese, Roma, 1955.

[4] G.FICHERA - Analisi esistenziale per le soluzioni dei problemi al contorno misti etc. - Annali Scuola Norm. Sup. Pisa, 1947.

[5] G.GIRAUD - Equations a intégrales principales - Ann. Ec. Norm. Sup. t. 51, 1934.

[6] G.GIRAUD - Problèmes mixtes et problèmes sur des varietés closes etc. - Ann. soc. Polon. Mathem., 1932.

[7] C.MIRANDA - Equazioni alle derivate parziali di tipo ellittico - Ergeb. Springer, 1955.

[8] C.MIRANDA - Sul problema misto per le equazioni lineari ellittiche- Ann. di Matem. , 1955.

[9] G.STAMPACCHIA - Problemi al contorno ellittici con dati discontinui, dotati di soluzioni hölderiane - Ann. di Matem., 1960.

Lecture 12

The classical elliptic BVP of Mathematical Physics: Linear Elastostatics.

We shall now consider the classical BVP of linear elasticity in the case of an inhomogeneous anisotropic elastic body. It is convenient to study the BVP connected with the equilibrium problems in the space X^r in order to include both the cases of plane and 3-dimensional elasticity. Set:

$$\varepsilon_{ih} = 2^{-1}\left(\frac{\partial u_i}{\partial x_h} + \frac{\partial u_h}{\partial x_i}\right) \qquad (i,h = 1,\dots,r).$$

Let us consider the elastic potential

$$W(x,\varepsilon) = \sum_{i\le h}^{1,r} \sum_{j\le k}^{1,r} a_{ih,jk}(x)\,\varepsilon_{ih}\,\varepsilon_{jk}\,.$$

The (real-valued) functions $a_{ih,jk}(x)$ are supposed to belong to C^∞ and the quadratic form $W(x,\varepsilon)$ is supposed to be positive definite in the $\frac{r(r+1)}{2}$ variables ε_{ih} $(1\le i\le h\le r)$ for any $x\in X^r$. We can suppose that:

$$a_{ih,jk}(x) \equiv a_{jk,ih}(x).$$

Let us now define for $i, h, j, k,$ arbitrary values of the indeces $1, \dots, r$:

$$
a_{ih,jk}(x) \begin{cases} = \alpha_{hi,jk}(x) & \text{for } i > h,\ j \leq k \\ = \alpha_{ih,kj}(x) & \text{for } i \leq h,\ j > k \\ = \alpha_{hi,kj}(x) & \text{for } i > h,\ j > k \\ = 2\alpha_{ih,jk}(x) & \text{for } i = h,\ j = k. \end{cases}
$$

We have for the elastic potential:

$$
(12.1) \quad W(x,\varepsilon) = \frac{1}{2} a_{ih,jk}\, \varepsilon_{ih}\, \varepsilon_{jk} = \frac{1}{2} a_{ih,jk} \frac{\partial u_i}{\partial x_h} \frac{\partial u_j}{\partial x_k} .
$$

It must be pointed out that the quadratic form $a_{ih,jk}\,\eta_{ih}\,\eta_{ik}$ is not positive definite, but only semidefinite as a function of the r^2 real variables η_{ih} $(i,h = 1,\dots,r)$. It is positive definite only in the subspace of the r^2-dimensional space of the η_{ih}'s defined by the conditions $\eta_{ih} = \eta_{hi}$.

Let A be the C^∞-smooth domain considered in lecture 11. The equations of equilibrium in A are the following:

$$
(12.2) \qquad \frac{\partial}{\partial x_h} \frac{\partial}{\partial \varepsilon_{ih}} W(x,\varepsilon) + f_i = 0 \qquad \text{in } A.
$$

We have three kinds of boundary conditions corresponding to the three main problems of elasticity. We consider here only homogeneous boundary conditions.

<u>1st BVP</u> (body fixed along its boundary)

(12.3) $$u = 0 \qquad \text{on} \quad \partial A .$$

<u>2nd BVP</u> (body free along its boundary)

(12.4) $$t_i(u) \equiv \nu_h \frac{\partial}{\partial \varepsilon_{ih}} W(x, \varepsilon) = 0 \quad \text{on} \quad \partial A .$$

(ν is the unit inward normal to ∂A)

<u>3rd BVP</u> (mixed BVP)

(12.5) $u = 0$ on $\overline{\partial_1 A}$, (12.6) $t(u) = 0$ on $\partial_2 A$,

where $\partial_1 A$ and $\partial_2 A$ are the subsets of ∂A already introduced in lecture 11.

Other BVP's could be considered, for instance the ones assigning p components of u and $n-p$ components of $t(u)$ on ∂A. However we shall restrict ourselves to the three above considered cases and leave it as an exercise to the reader to study other BVP's for (12.2). Equations (12.2) can be written:

(12.2') $$\frac{\partial}{\partial x_h} \left[a_{ih,jk}(x) \frac{\partial u_j}{\partial x_k} \right] + f_i = 0 .$$

In order to prove the existence and the uniqueness of the solution of BVP (12.2) (12.3), we need to prove inequality (9.1), which in the present case is the following:

(12.7) $$\int_A a_{ih,jk}(x)\,\frac{\partial v_i}{\partial x_h}\,\frac{\partial v_j}{\partial x_k}\,dx \geq c_0 \|v\|_1^2$$

for any $v \in \overset{\circ}{H}_1(A)$. Because of (12.1) inequality (12.7) - which is known as the <u>1st Korn's inequality</u> - is equivalent to the following one:

(12.8) $$\int_A \sum_{i,h}^{1,r} \left(\frac{\partial v_i}{\partial x_h} + \frac{\partial v_h}{\partial x_i}\right)^2 dx \geq c_1 \|v\|_1^2 \qquad (c_1 > 0).$$

This is immediately obtained by using the Fourier developments of the functions v_i and Parseval's theorem.

It must be observed that, as a consequence of (12.7), it follows that for any $x \in \bar{A}$ and any non-zero real ξ :

$$a_{ih,jk}(x)\,\xi_h\,\xi_k\,\eta_i\,\eta_j > 0$$

for every non-zero real r-vector η. This is a consequence of a general theorem which will be proved later (see theor. 14.II of lecture 14). In particular, the ellipticity of system (12.2) follows. We have thus the following theorem.

12.I. <u>Given</u> $f \in C^\infty(\bar{A})$, <u>there exists one and only one solution of the BVP (12.2), (12.3), which belongs to</u> $C^\infty(\bar{A})$.

In order to prove the existence theorem for (12.2), (12.4), let us consider the system:

(12.9) $$\frac{\partial}{\partial x_h}\left[a_{ih,jk}(x)\,\frac{\partial u_j}{\partial x_k}\right] - p_i u_i + f_i = 0$$

where p_0 is any positive constant. We wish first solve problem (12.9) (12.4). It is easily seen that the inequality to be proven in this case is the following (2nd Korn's inequality[1])

$$\text{(12.10)} \qquad \int_A \sum_{i,h}^{1,n} \left(\frac{\partial v_i}{\partial x_h} + \frac{\partial v_h}{\partial x_i}\right)^2 dx + \int_A |v|^2 dx \geq c_2 \| v \|_1^2$$

for any $v \in H_1(A)$. Several proofs of (12.10) have been given after the original one due to Korn [5] (see [2],[6],[3]). A rather simple one can be obtained (see [6]) if $\bar{A}$ is C^∞-homeomorphic to a closed ball. We refer the reader to the quoted papers.

From (12.10) it follows that (12.9) (12.4) has only one solution which is C^∞ in $\bar{A}$. Since our differential system is formally self-adjoint, it follows that a C^∞ solution of the following differential system:

$$\text{(12.11)} \qquad \frac{\partial}{\partial x_h}\left[a_{ih,jk}(x)\frac{\partial u_j}{\partial x_k}\right] - p_0 u_i + \lambda u_i + f_i = 0,$$

with the boundary conditions (12.4), exists when and only when:

$$\text{(12.12)} \qquad \int_A f_i \rho_i \, dx = 0 ;$$

(1) Actually the 2nd Korn's inequality is the following:

$$\int_A \sum_{ih}^{1,n} \left(\frac{\partial v_i}{\partial x_h} + \frac{\partial v_h}{\partial x_i}\right)^2 dx \geq c_3 \int_A \sum_{ih}^{1,n} \left(\frac{\partial v_i}{\partial x_h}\right)^2 dx \qquad (c_3 > 0)$$

for any v such that:

$$\int_A \left(\frac{\partial v_i}{\partial x_h} - \frac{\partial v_h}{\partial x_i}\right) dx = 0 .$$

However it is easily seen that this inequality implies and is implied by (12.10).

ρ_i is any C^∞ solution of (12.11) (12.4) with $f_i \equiv 0$. In the case $\lambda = p_0$ the only C^∞ solution of the homogeneous system is :

(12.13) $$\rho_i = a_i + b_{ij} x_j$$

where a_i and b_{ij} are arbitrary constants such that $b_{ij} = -b_{ji}$.

12.II. The BVP (12.2) (12.4) has solutions belongings to $C^\infty(\bar{A})$ if and only if the C^∞ function f satisfies conditions (12.12) with ρ_i given by (12.13).

For getting the existence and uniqueness of the solution of the BVP (12.2), (12.5), (12.6) we assume that V is the subspace of $H_1(A)$ composed of the functions vanishing on $\partial_1 A$. From the second Korn's inequality (considered for any $v \in V$) it is easy to derive, for any $v \in V$, inequality (12.8). Arguing as in the previous lecture for the case of the mixed BVP for a 2nd order elliptic equation, we deduce the following theorem:

12.III. For $f \in C^\infty(\bar{A})$ there exists one and only one solution of the BVP (12.2) (12.5) (12.6), which belongs to $C^\infty(A \cup \partial_1 A \cup \partial_2 A) \cap H_1(A)$.

It is worthwhile to remark that the obtained solutions of the 1st, 2nd and 3rd BVP's are the ones required by the mathematical theory of elasticity, since they minimize the energy integral

$$\int_A [W(x,\varepsilon) - uf]\, dx$$

in the classes $\mathring{H}_1(A)$, $H_1(A)$ and V respectively.

Bibliography of Lecture 12.

[1] G. FICHERA - see [2] of lecture 4.

[2] K.O.FRIEDRICHS - On the Boundary-value Problems of the Theory of Elasticity and Korn's inequality - Annals of Math. v. 48, 1947.

[3] J. GOBERT - Une inégalité fondamentale de la théorie de l'élasticité - Bull. Soc. Roy. des Sci. de Liège - 3-4 - 1962.

[4] A. KORN - Solution generale du probleme d'equilibre dans la theorie de l'élasticité dans le cas ou les efforts sont donnés à la surface - Ann. Toulose Univ. 1908.

[5] A. KORN - Ueber einige Ungleichungen welche in der Theorie der elastischen und elektrischen Schwingungen eine Rolle spielen - Bull. Inst. Cracovie Akad. Umiejet, Classe des sci. math. et nat. , 1909.

[6] L.E.PAYNE-H.F.WEINBERGER - On Korn's Inequality - Arch. for Rat. Mech. & Anal. 8, 1961.

Lecture 13

The classical elliptic BVP of Mathematical Physics: Equilibrium of thin plates.

The classical theory of the equilibrium of thin plates requires the solution of certain BVP's for the iterated Laplace operator in two variables x, y :

$$\Delta_4 u = \frac{\partial^4 u}{\partial x^4} + 2\frac{\partial^4 u}{\partial x^2 \partial y^2} + \frac{\partial^4 u}{\partial y^4}$$

with several kinds of boundary conditions. Let us suppose that A is a C^∞-smooth bounded (connected) plane domain. The theory of thin plates considers the following boundary conditions on ∂A

(13.1) $\quad u = 0,$ $\qquad$ (13.2) $\quad \frac{\partial u}{\partial \nu} = 0,$

(13.3) $$\frac{\partial^2 u}{\partial \nu^2} + \sigma\left(\frac{\partial^2 u}{\partial s^2} - \frac{1}{\rho}\frac{\partial u}{\partial \nu}\right) = 0,$$

(13.4) $$\frac{\partial}{\partial \nu}\Delta_2 u + (1-\sigma)\frac{\partial}{\partial s}\frac{\partial^2 u}{\partial \nu \partial s} = 0 .$$

Here ν is the unit inward normal to ∂A ; $\frac{\partial}{\partial s}$ denotes differentiation with respect to the arc (increasing counter-clockwise) ; $\frac{1}{\rho}$ is the curvature of ∂A ; σ is a constant such that $-1<\sigma<1$; Δ_2 is the Laplace operator.

The differential equation to be considered is the following (u and f real valued functions):

(13.5) $$\Delta_4 u = f .$$

The BVP (13.5), (13.1), (13.2) corresponds to the equilibrium problem for a plate clamped along its boundary. The boundary conditions (13.1), (13.3) express the fact that the plate is supported along its edge. The boundary conditions (13.3) and (13.4) mean that the part of the boundary where these conditions are satisfied is free.

We restrict ourself to the consideration of the BVP (13.5),(13.1), (13.2) and the mixedBVP for a partially clamped plate, i.e. when (13.1) and (13.2) are satisfied on a part $\overline{\partial_1 A}$ of ∂A and (13.3), (13.4) on the remaining part. $\partial_1 A$ and $\partial_2 A$ are the subsets of ∂A considered in lecture 11. The reader is requested to carry out the proofs of the existence and uniqueness theorems in the other cases, for instance in the BVP corresponding to a plate partially clamped on ∂A , partially supported and partially free.

As bilinear form $B(u,v)$ corresponding to the BVP (13.5),(13.1), (13.2) we assume:

$$B(u,v) = \int_A \left(\frac{\partial^2 u}{\partial x^2} \frac{\partial^2 v}{\partial x^2} + 2 \frac{\partial^2 u}{\partial x \partial y} \frac{\partial^2 v}{\partial x \partial y} + \frac{\partial^2 u}{\partial y^2} \frac{\partial^2 v}{\partial y^2} \right) dx\, dy .$$

The subspace V of $H_2(A)$ to be considered is $\mathring{H}_2(A)$. In this case inequality (9.1) reduces to inequlity (3.6) for $m=2$. Thus we have

$$B(v,v) \geq c_0 \|v\|_2^2 . \qquad [v \in \mathring{H}_2(A)]$$

Then:

13.I. <u>If</u> $f \in C^\infty(\bar{A})$, <u>BVP (13.5), (13.1), (13.2) has one and only one solution belonging to</u> $C^\infty(\bar{A})$.

In order to consider the above mentioned mixed BVP (i.e. conditions (13.1),(13.2) on $\partial_1 A$ and conditions (13.3),(13.4) on $\partial_2 A$) it is convenient to observe that, for u and v belonging to $C^\infty(\bar{A})$, we have:

$$\int_{\partial A} v \left[\frac{\partial}{\partial \nu} \Delta_2 u + (1-\sigma) \frac{\partial}{\partial s} \frac{\partial^2 u}{\partial \nu \partial s} \right] ds -$$

$$- \int_{\partial A} \frac{\partial v}{\partial \nu} \left[\frac{\partial^2 u}{\partial \nu^2} + \sigma \left(\frac{\partial^2 u}{\partial s^2} - \frac{1}{\rho} \frac{\partial u}{\partial \nu} \right) \right] ds =$$

(13.6)

$$= \int_A \left\{ \frac{\partial^2 u}{\partial x^2} \frac{\partial^2 v}{\partial x^2} + (2-2\sigma) \frac{\partial^2 u}{\partial x \partial y} \frac{\partial^2 v}{\partial x \partial y} + \frac{\partial^2 u}{\partial y^2} \frac{\partial^2 v}{\partial y^2} + \right.$$

$$\left. + \sigma \left[\frac{\partial^2 u}{\partial x^2} \frac{\partial^2 v}{\partial y^2} + \frac{\partial^2 u}{\partial y^2} \frac{\partial^2 v}{\partial x^2} \right] \right\} dx\, dy - \int_A v \, \Delta_4 u \; dx\, dy.$$

Let us now assume as space V the subspace of $H_2(A)$ composed of the functions which satisfy conditions (13.1),(13.2) on $\partial_1 A$. Set

$$B(u,v) = \int_A \left\{ \frac{\partial^2 u}{\partial x^2} \frac{\partial^2 v}{\partial x^2} + (2-2\sigma) \frac{\partial^2 u}{\partial x \partial y} \frac{\partial^2 v}{\partial x \partial y} + \right.$$

$$\left. + \frac{\partial^2 u}{\partial y^2} \frac{\partial^2 v}{\partial y^2} + \sigma \left[\frac{\partial^2 u}{\partial x^2} \frac{\partial^2 v}{\partial y^2} + \frac{\partial^2 u}{\partial y^2} \frac{\partial^2 v}{\partial x^2} \right] \right\} dx\, dy.$$

Because of the assumption $-1 < \sigma < 1$, we have:

$$B(v,v) \geq c(\sigma) \sum_{|p|=2} \int_A |D^p v|^2 \, dx\, dy.$$

The constant $c(\sigma)$ depends only on σ.

In order to prove (9.1) we need only to show that there exists $c_1 > 0$ such that, for any $v \in V$

$$\sum_{|p|=2} \int_A |D^p v|^2 \, dx\, dy \geq c_1 \|v\|_2^2 . \tag{13.7}$$

Suppose (13.7) to be false. Then there exists $\{v_m\} \in V$ such that:

(13.8) $\|v_m\|_2 = 1$, (13.9) $\displaystyle\sum_{|p|=2} \int_A |D^p v_m|^2 \, dx\, dy < \frac{1}{m}$.

We can suppose that $\{v_n\}$ converges in $H_1(A)$ (theor. 3.IV). Then, for (13.9), $\{v_n\}$ converges in $H_2(A)$ and the limit function v has strong second derivatives vanishing on A. It follows readily that v is a polynomial of degree 1. Because v belongs to V, $v \equiv 0$ in A. This contradicts (13.8).

Since (9.1) has been proved, there exists one and only one solution u of the equations:

$$B(u,v) = \int_A v f \, dx \qquad (v \in V)$$

belonging to V. The function u belongs to $C^\infty(A \cup \partial_1 A \cup \partial_2 A)$. By using (13.6), it follows that u is the solution of our mixed problem.

13.II. <u>The mixed BVP (13.5); (13.1),(13.2) on</u> $\partial_1 A$; <u>(13.3), (13.4) on</u> $\partial_2 A$, $f \in C^\infty(\bar{A})$, <u>has one and only one solution belonging to</u> $C^\infty(A \cup \partial_1 A \cup \partial_2 A) \cap H_2(A)$.

From lemma 4.II it follows also that u belongs to $C^0(\bar{A})$.

<u>Bibliography of Lecture 13</u>

[1] G. FICHERA - <u>Teorema d'esistenza per il problema bi-iperarmonico</u> - Rend. Acc. Naz. Lincei, 1948.

[2] G. FICHERA - <u>On some general integration methods employed in connection with linear differential equations</u> - Jour. of Math. and Phy. vol.XXIX , 1950.

[3] G. FICHERA - Esistenza del minimo in un classico problema di calcolo delle variazioni - Rend. Acc. Naz. Lincei, 1951.

[4] K.O.FRIEDRICHS - Rie Randwert- und Eigenwert Probleme aus der Theorie der elastischen Platten - Math. Annalen, 1928.

[5] G. FUBINI - Il principio di minimo e i teoremi di esistenza per i problemi al contorno relativi alle equazioni alle derivate parziali di ordine pari - Rend. Circ. Matem. Palermo, 1907.

[6] A.E.H.LOVE - A Treatise on the Mathematical Theory of Elasticity - Cambridge at the Univ. Press - vol.I, 1893.

Lecture 14

Strongly elliptic operators. Gårding inequality. Eigenvalue problems.

The existence theory developed in the previous lectures is founded on inequality (9.1). We proved this inequality for all of the particular cases considered in lectures 11, 12, and 13. We wish now consider a general operator. Of course the possibility of proving (9.1) depends on the choice of the subvariety V of the space $H_m(A)$. However, in any case, because of the assumption $\overset{\circ}{H}_m(A) \subset V$, inequality (9.1) must be true when $\overset{\circ}{H}_m(A) = V$. We shall consider here only this case, which corresponds to the Dirichlet problem.

The matrix differential operator $L \equiv a_s(x) D^s \quad (0 \leq |s| \leq \nu)$ is said to be a strongly elliptic operator at the point x, if for any real non-zero r-vector ξ and for every non-zero complex n-vector η we have:

$$\Re e \Big(\sum_{|s|=\nu} a_s(x) \xi^s \eta \Big) \eta \neq 0 . \text{ [1]}$$

It is evident that strong ellipticity implies ellipticity. The converse is not true, as the example of the Wirtinger operator $\frac{1}{2}\left(\frac{\partial}{\partial x_1} + i \frac{\partial}{\partial x_2}\right)$ proves.

---

[1] For a more general definition of strong ellipticity, see [4]. From now on we shall omit parentheses when writing $\big(\sum_{|s|=\nu} a_s(x)\xi^s \eta\big)\eta$.

For the operator $L \equiv D^p a_{pq}(x) D^q \quad (0 \leq |p| \leq m,\ 0 \leq |q| \leq m)$ we shall assume that the strong ellipticity condition holds at any point $x \in \bar{A}$, where A is a bounded domain of X^r. Since A is connected, we can suppose that the quadratic form in the complex vector η

$$Q(x, \xi, \eta) \equiv \mathcal{R}e \sum_{|p|,|q|=m} a_{pq}(x) \xi^p \xi^q \eta \bar{\eta}$$

is definite, for instance positive, for $x \in \bar{A}$ and any $\xi \neq 0$.

Let us suppose the coefficients $a_{pq}(x)$ $(|p|=|q|=m)$ of the operator L to be continuous functions in $\bar{A}$. If $p_0 > 0$ is the minimum of the positive function $Q(x, \xi, \lambda)$ on the compact set: $x \in \bar{A}$, $|\xi| = 1$, $|\eta| = 1$, we have for any ξ and any η:

$$\mathcal{R}e \sum_{|p|,|q|=m} a_{pq}(x) \xi^p \xi^q \eta \bar{\eta} \geq p_0 |\xi|^{2m} |\eta|^2. \tag{14.1}$$

We cannot expect that (9.1) holds for any L, even one that is strongly elliptic, since this would imply - for $V = \mathring{H}_m(A)$ - existence and uniqueness for the Dirichlet problem for the equation $Lu = f$. What we want to prove is that, for some λ, inequality (9.1) is true for the operator $L + \lambda I$ (I = identity).

We shall prove the following theorem concerning Gårding's inequality

14.I. There exist two positive constants γ_0 and λ_0 such that for any $v \in \mathring{H}_m(A)$ the following inequality holds:

$$(-1)^m \mathcal{R}e\, B(v,v) \geq \gamma_0 \|v\|_m^2 - \lambda_0 \|v\|_0^2 . \tag{14.2}$$

The coefficients $a_{pq}(x)$ are supposed continuous in $\bar{A}$ for $|p|=|q|=m$

<u>and measurable and bounded in</u> A <u>for</u> $|p|+|q|<2m$.

Set

$$\beta_{ls}=(-1)^{|l|+m}a_{ls} \qquad \text{for} \qquad |l|+|s|<2m.$$

Then we can write:

$$(-1)^m B(u,v)=\int_A a_{pq}D^q u\, D^p v\, dx+\int_A \beta_{ls}D^s u\, D^l v\, dx.$$

In the first integral the summation must be understood to be restricted to $|p|=|q|=m$, in the second to $|l|+|s|<2m$, $0\le |l|\le m$, $0\le |s|\le m$. From now on, in the course of this proof, we write a_{pq} only in the case where $|p|=|q|=m$.

The proof is obtained by considering three cases.

<u>1st case:</u> a_{pq} = constant, $\beta_{ls}\equiv 0$.

We suppose, without any loss in generality, that $\bar{A}\subset Q$ ($Q: |x_h|<\pi$, $h=1,\dots,r$).

By using (3.3) and (3.4), we have for $v\in \overset{\circ}{H}_m(A)$

$$D^p v=\frac{1}{(\sqrt{2\pi})^r}\sum_{-\infty}^{+\infty}{}_{\kappa}\, i^{|p|}\kappa^p c_\kappa e^{i\kappa x}$$

$$a_{pq}D^q v=\frac{1}{(\sqrt{2\pi})^r}\sum_{-\infty}^{+\infty}{}_{\kappa}\, i^{|q|}\kappa^q a_{pq}c_\kappa e^{i\kappa x}.$$

Then, by Parseval's theorem, (14.1), and theor. 3.II,

$$(-1)^m \mathcal{R}e\, B(v,v) = \sum_{-\infty}^{+\infty}{}_{\kappa}\, \mathcal{R}e\, a_{pq}\, \kappa^p \kappa^q c_\kappa \bar{c}_\kappa \geq$$

$$\geq p_0 \sum_{-\infty}^{+\infty} |\kappa|^{2m} |c_\kappa|^2 \geq \tilde{\gamma}_0 \|v\|_m^2 .$$

In this particular case (14.2) holds with $\gamma_0 = \tilde{\gamma}_0$ and $\lambda_0 = 0$.

<u>2nd case:</u> diameter spt $v < \delta_0$ (where δ_0 is a proper positive number).

Let $x^0 \in$ spt v . We have:

$$\tilde{\gamma}_0 \|v\|_m^2 \leq \mathcal{R}e \int_A a_{pq}(x^0) D^q v \overline{D^p v}\, dx = \mathcal{R}e \int_A a_{pq}(x) D^q v \overline{D^p v}\, dx +$$

$$+ \mathcal{R}e \int_A \beta_{l,s}(x) D^s v \overline{D^l v}\, dx + \mathcal{R}e \int_A [a_{pq}(x^0) - a_{pq}(x)] D^q v \overline{D^p v}\, dx -$$

$$- \mathcal{R}e \int_A \beta_{ls}(x) D^s v \overline{D^l v}\, dx \leq (-1)^m \mathcal{R}e\, B(v,v) +$$

(2)

$$+ \max_{x \in \text{spt}\, v} \sum_{p,q} |a_{pq}(x^0) - a_{pq}(x)| \, \|v\|_m^2 + \gamma_1 \|v\|_m \|v\|_{m-1} .$$

---

(2) During this proof and the proof of the next theorem, $\gamma_1, \gamma_2, \gamma_3, \ldots$ denote positive constants"

Let diam. spt $v < \delta_o$, where δ_o is so small that:

$$\max_{x \in \operatorname{spt} v} \sum_{p,q} |a_{pq}(x^o) - a_{pq}(x)| \leq \frac{1}{2}\tilde{\gamma}_o .$$

Then we have:

$$\frac{1}{2}\tilde{\gamma}_o \|v\|_m^2 \leq (-1)^m \operatorname{Re} B(v,v) + \gamma_1 \|v\|_m \|v\|_{m-1} .$$

From this it follows that:

$$\|v\|_m^2 \leq \gamma_2 (-1)^m \operatorname{Re} B(v,v) + \gamma_3 \|v\|_{m-1}^2 .$$

From lemma 4.III, inequality (14.2) follows.

<u>3rd case:</u> general case.

Let us consider the following partition of unity in $\bar{A}$: $\sum_{h=1}^{\nu} \varphi_h^2(x) = 1$ with $\varphi_h(x) \in C^\infty$ and diam spt $\varphi_h(x) < \delta_o$, where δ_o is such that for $|x^1 - x^2| < \delta_o$ one has:

$$\sum_{pq} |a_{pq}(x^1) - a_{pq}(x^2)| < \frac{1}{2}\tilde{\gamma}_o .$$

We have:

$$(-1)^m B(v,v) = \sum_{h=1}^{\nu} \int_A \varphi_h^2(x)\, a_{pq} D^q v D^p v\, dx + \int_A \beta_{ls} D^s v D^l v\, dx =$$

$$= \sum_{h=1}^{\nu} \int_A a_{pq} D^q \varphi_h v\, D^p \varphi_h v\, dx + O(\|v\|_m \|v\|_{m-1}) .$$

From the 2nd case considered above we have:

$$\|\varphi_h v\|_m^2 \leq \gamma_5 \, \mathrm{Re} \int_A a_{pq} D^q \varphi_h v \, \overline{D^p \varphi_h v} \, dx + \gamma_6 \|\varphi_h v\|_0^2 .$$

Thus,

$$(-1)^m \, \mathrm{Re}\, B(v,v) \geq \gamma_7 \|v\|_m^2 - \gamma_8 \|v\|_0^2 - \gamma_9 \|v\|_m \|v\|_{m-1}$$

From this inequality, by using the same argument as in the 2nd case, we see that (14.2) follows.

<u>Remark</u>. Retracing the steps of the proof of (14.2) we see that γ_0 and λ_0 depend on ρ_0, on the modulus of continuity of a_{pq}, on an upper bound for $|a_{pq}|$ and $|\beta_{ls}|$ and on the domain A. It is possible to express explicitly γ_0 and λ_0 in terms of these objects.

14.II. <u>Under the same hypotheses on</u> a_{pq} $(|p| = |q| = m)$ <u>and on</u> β_{ls} <u>as in the previous theorem, if (14.2) holds for any</u> $v \in \overset{\circ}{H}_m(A)$, <u>then the operator</u> L <u>is strongly elliptic in</u> $\bar{A}$.
Suppose the theorem is not true. Then (14.2) holds and there exists a real $\xi \neq 0$ and a complex $\eta \neq 0$ such that for some $x^0 \in A$ one has

$$(14.3) \qquad \mathrm{Re} \sum_{|p|,|q|=m} a_{pq}(x^0) \xi^p \xi^q \eta \bar{\eta} = 0.$$

For $v \in \overset{\circ}{H}_m(A)$,

$$\gamma_0 \|v\|_m^2 \leq (-1)^m \operatorname{Re} B(v,v) + \lambda_0 \|v\|_0^2 \leq \operatorname{Re} \int_A a_{pq}(x^0) D^q v \, D^p v \, dx +$$

(14.4)

$$+ \operatorname{Re} \int_A [a_{pq}(x) - a_{pq}(x^0)] D^q v \, D^p v \, dx + \gamma_1 \|v\|_m \|v\|_{m-1} .$$

Let δ_0 **be such that for** $|x - x_0| < \delta_0$:

(14.5) $$\sum_{p,q} |a_{pq}(x) - a_{pq}(x^0)| < \frac{1}{2} \gamma_0 .$$

Let I_0 **be the ball** $|x - x^0| < \delta_0$. **If** $\operatorname{spt} v \subset A \cap I_0$ **, then from (14.4), (14.5) and lemma 4.III, we deduce that:**

(14.6) $$\|v\|_m^2 \leq \gamma_{10} \operatorname{Re} \int_A a_{pq}(x^0) D^q v \, D^p v \, dx + \gamma_{11} \|v\|_0^2 .$$

Let $\varphi(x)$ **be a** C^∞ **real valued function which is not identically zero and which has its support in** $A \cap I_0$. **Set:**

$$v = \varphi(x) e^{i\lambda \xi x} \eta$$

with λ **real. We have:**

$$\|v\|_m^2 = \lambda^{2m} |\eta|^2 \sum_{|p|=m} (\xi^p)^2 \int_A |\varphi(x)|^2 dx + O(\lambda^{2m-1}),$$

$$\operatorname{Re} \int_A a_{pq}(x^0) D^q v \, D^p v \, dx = \lambda^{2m} \operatorname{Re} a_{pq}(x^0) \xi^p \xi^q \eta \eta \int_A |\varphi(x)|^2 dx + O(\lambda^{2m-1}).$$

Because of (14.3), the right hand side of (14.6) is $O(\lambda^{2m-1})$. This disproves (14.6). Thus we have reached an absurdity.

Let us now consider any subspace V of $H_m(A)$, containing $\mathring{H}_m(A)$, such that (14.2) holds for any $v \in V$. Let us consider the operator $L_0 = L + (-1)^m \lambda_0 I$. The corresponding bilinear form is:

$$B_0(u,v) = B(u,v) + (-1)^m \lambda_0 (u,v)_0 .$$

From (14.2) it follows that for $v \in V$,

(14.7) $$|\mathrm{Re}\, B_0(v,v)| \geq \gamma_0 \|v\|_m^2 .$$

Then there exists one and only one $u \in V$ such that:

(14.8) $$B_0(u,v) = \int_A f v \, dx \qquad (v \in V)$$

for any $f \in \mathcal{L}^2(A)$. Let us denote by $u = Gf$ the solution of (14.8). G is the Green's transformation introduced earlier. However we do not want now to consider any result arising from the regularisation theory.

From (7.6) it follows that G is a bounded linear transformation from $\mathcal{L}^2(A)$ into V. Then, if considered as a linear operator from $\mathcal{L}^2(A)$ into itself, G is compact due to Rellich's lemma.

We wish to give some more information concerning the functional structure of the operator G.

Let us introduce the bounded linear operator T of the space V which was already considered in lecture 7, and for which $B_0(u,v) =$

$=(u,Tv)_m$. For any $v \in V$, inequality (7.8) is satisfied. Hence $T(V)$ is closed and T maps V in a one-to-one fashion onto $T(V)$. On the other hand, if w is orthogonal to $T(V)$, then, from (14.7), it follows that $w=0$. Thus $T(V)=V$. Let $\mathcal{G}^*$ be the inverse operator of T; $\mathcal{G}^*$ is bounded because of (7.8). So is the adjoint operator $\mathcal{G}$ of $\mathcal{G}^*$. We have for any $h \in V$: $(\mathcal{G}h, Tv)_m = (h,v)_m$. Let $\mathfrak{I}_{mo}$ be the embedding of $H_m(A)$ into $H_o(A)$, and $\mathfrak{I}^*_{mo}$ denote its adjoint operator, i.e. an operator from H_o into H_m such that for $f \in H_o$ and $v \in H_m$, $(f, \mathfrak{I}_{mo} v)_o = (\mathfrak{I}^*_{mo} f, v)_m$. It follows that the operator G, thought as an operator of $\mathcal{L}^2(A)$, is given by:

$$G = \mathfrak{I}_{mo}\, \mathcal{G}\, \mathfrak{I}^*_{mo}.$$

If L is formally self-adjoint, i.e. $L \equiv L^*$ (3), then T is hermitian and so are $\mathcal{G}$ and G. It is worthwhile to remark that $\mathfrak{I}^*_{mo}$ is itself a kind of Green's operator for a differential operator with constant coefficients, whose leading part, after inessential changes in the definition of $(\ ,\ )_m$, is the iterated Laplace operator Δ_2^m.

We want now to consider the eigenvalue problem:

$$B_o(u,v) - (-1)^m \lambda \int_A u v \, dx = 0 \qquad (v \in V). \tag{14.9}$$

When the results of the regularisation theory can be applied, then (14.9) corresponds to the eigenvalue problems for the differential operator

$$L_o u - (-1)^m \lambda u = 0 \qquad , \quad u \in M(V).$$

(3) It must be pointed out that when we write $L \equiv L^*$, we mean $a_{pq} = (-1)^{|p|+|q|} \bar{a}_{qp}$ i.e. $B(u,v)$ is hermitian (see also pag.76).

In any case, (14.9) is equivalent to the following eigenvalue problem in the space $\mathcal{L}^2(A)$:

(14.10) $$\varphi - (-1)^m \lambda G\varphi = 0.$$

Suppose that L is self-adjoint. Then from the Hilbert space theory of eigenvalue problems for compact hermitian operators it follows that there exists a countable set of real eigenvalues for problem (14.9). These eigenvalues are positive since the operator G is a positive operator. The constant γ_0 is a lower bound for the eigenvalues of problem (14.9).

In the next lecture we shall develop a theory for computing lower and upper bounds for the eigenvalues of problems which include, as a particular case, problem (14.10).

Remark. The way as we derived the structure of the Green's transformation G suggests a more abstract approach to elliptic problems. In fact, the differential structure of the operator L is only needed in order to consider the integro-differential form $B_0(u,v)$. However, what is only needed in the existence theory is that $B_0(u,v)$ be continuous over $V \times V$, i.e. that $B_0(u,v) = (u,Tv)_m$ with T linear and bounded. The question naturally arises of considering more general bilinear forms such that (14.7) holds and such that the regularisation theory can still be carried out.

For a similar approach in the theory of harmonic differential forms on a Riemannian compact manifold, see [2].

Bibliography of Lecture 14.

[1] N. ARONSZAJN - On coercive integro-differential quadratic forms - Conference on PDE, Univ. of Kansas 1954, Tech. Rep. No 14.

[2] G. FICHERA - Teoria assiomatica delle forme differenziali armoniche, Rend. di Matem., Roma vol.20, 1961.

[3] L. GÅRDING - Dirichlet's problem for linear elliptic partial differential equation, Math. Scand. vol.I, 1953.

[4] L. NIRENBERG - see [4] of lecture 4.

[5] L. NIRENBERG - see [4] of lecture 8.

Lecture 15

Eigenvalue Problems . The Rayleigh-Ritz method.

We assume that the reader is acquainted with the elementary theory of hermitian compact (i.e. completely continuous) operators in a Hilbert space. However we shall list here some definitions and theorems which we shall need in the sequel. For the proofs of the theorems we refer to [2],[3],[4].

We shall always consider operators which are linear and are defined in the whole of the complex Hilbert space S . An operator T is said to be <u>hermitian</u> if $(Tu,v) = (u,Tv)$ for any pair of vectors u and v of S .

15.I. <u>Any hermitian operator is bounded</u>.

The operator T is <u>positive</u> if $(Tu,u) \geq 0$ for any $u \in S$.

15.II. <u>A (linear) positive operator is hermitian</u>.

15.III. <u>A (linear) compact operator is bounded</u>.

T is said to be <u>strictly positive</u> if it is positive and $(Tu,u) = 0$ only holds for $u = 0$.

We write PCO for any positive compact operator. A PCO has a sequence of positive eigenvalues converging towards zero. Each positive eigenvalue has finite (geometric) multiplicity, i.e., if μ is a positive eigenvalue for T , the subspace of S composed of all the solutions of the homogeneous equation $Tu - \mu u = 0$ is finite dimensional.

15.IV. A PCO T is strictly positive if and only if $\mu = 0$ is not an eigenvalue for T.

Let:

$$\mu_1 \geq \mu_2 \geq \cdots \geq \mu_k \geq \cdot$$

be the sequence of all the eigenvalues of the PCO T each repeated as many times as its multiplicity. From now on when we mention the sequence of eigenvalues of a PCO, it shall be understood that this sequence is ordered according to the criterion just now specified. Let:

$$u_1, u_2, \ldots, u_k, \ldots$$

be a corresponding sequence of eigenvectors, i.e. $Tu_k - \mu_k u_k = 0$. We may assume that $\{u_k\}$ is an orthonormal system.

15.V. The operator T admits the following "spectral decomposition":

$$Tu = \sum_k \mu_k (u, u_k) u_k . \tag{15.1}$$

15.VI. The system $\{u_k\}$ is a complete system in the space S, if and only if the PCO T is strictly positive.

It follows that a strictly positive compact operator can exist only when S is separable. From now on we shall assume this hypothesis on S and, moreover, that S has infinite dimension.

15.VII. If $\{\mu_k\}$ is any non-increasing sequence of non-negative real numbers converging towards zero and $\{u_k\}$ is any orthogonal system, then the operator defined by (15.1) is a PCO. having the μ_k as eigenvalues and the u_k as eigenvectors. The value $\mu = 0$ is an eigenvalue for T - even though it is not included in the sequence $\{\mu_K\}$ - if and only if the system of eigenvalues corresponding to all positive eigenvalues is not complete.

The following lemma is of fundamental interest in the theory of eigenvalues. It expresses the so-called mini-max property for eigenvalues.

15.VIII. Let T be the above considered PCO. Let $v_1, \ldots, v_{K-1}$ ($K \geq 1$) be $K-1$ vectors of the Hilbert space S. Denote by $M(v_1, \ldots, v_{K-1})$ the maximum of (Tu, u) on the unit sphere $\|u\| = 1$ of the orthogonal complement $S \ominus V_{K-1}$ of V_{K-1}, where V_{K-1} is the linear variety spanned by $v_1, \ldots, v_{K-1}$ (1). The minimum of $M(v_1, \ldots, v_{K-1})$ when $v_1, \ldots, v_{K-1}$ vary arbitrarily in the space S is the eigenvalue μ_K of T. (2) If $v_1 = u_1, \ldots, v_{K-1} = u_{K-1}$, then $M(v_1, \ldots, v_{K-1}) = \mu_K$.

The sequence of operators $\{T_n\}$ is said to converge weakly (strongly, uniformly) to the operator T whenever:

$$\lim_{n \to \infty} (T_n u, v) = (Tu, v) \qquad \text{for any } u, v \text{ in } S$$

$$\left[\lim_{n \to \infty} \|T_n u - Tu\| = 0 \qquad \text{for any } u \in S, \right.$$

$$\left. \lim_{n \to \infty} \|T_n - T\| = 0 \right].$$

In the latter case, the norm must be understood to be the norm in the space of operators. Therefore T_n and T are now supposed bounded.

---

(1) In the case $K = 1$, we assume $S \ominus V_{K-1} \equiv S$.

(2) Characterization of all possible choices of $v_1, \ldots, v_{K-1}$ such that $M(v_1, \ldots, v_{K-1}) = \mu_K$ has been done by Weinstein [5].

For weak, strong, uniform convergence of operators we shall use respectively the symbols:

$$T_n \rightharpoonup T \;,\quad T_n \longrightarrow T \;,\quad T_n \rightrightarrows T.$$

15.IX. Let T_n be a sequence of PCO's converging uniformly to the PCO T . Let $\{\mu_{nk}\}$ be the sequence of eigenvalues of T_n. We have:

$$\lim_{n\to\infty} \mu_{nk} = \mu_k$$

uniformly with respect to k .

The proof is a consequence of the interesting inequality:

$$|\mu_{nk} - \mu_k| \leq \|T_n - T\|,$$

which is proven as a consequence of the mini-max principle.

If V is a subspace (i.e. closed linear subvariety of S), then for any given vector $u \in S$ there exists one and only one vector of the subspace V which has minimum distance from u (projection theorem). This vector is characterised as the one satisfying the equations $(v, w) = (u, w)$ for every $w \in V$. The vector $v = Pu$ is called the (orthogonal) projection of u on V . The operator P which maps u onto its projection on V is called an (orthogonal) projector. A projector is characterised by the following two properties: i) P is hermitian ; ii) P is idempotent, i.e. $P^2 = P$.

Suppose V is finite dimensional and $v_1, \dots, v_m$ is a basis for V . Let $\{\alpha_{ih}\}$ be the inverse matrix of the Gramian matrix $\{v_h, v_k\}$ ($h,k = 1,\dots,m$), i.e. $\alpha_{ih}(v_h, v_k) = \delta_{ik}$. Since $(v_h, v_k) = (\overline{v_k, v_h})$,

$\alpha_{ih} = \bar{\alpha}_{hi}$. The projector P of S onto V has the following representation:

$$P u = \alpha_{ih} (u, v_i) v_h . \tag{15.2}$$

If the basis is orthonormal, then

$$P u = (u, v_i) v_i . \tag{15.3}$$

If V is any subspace of S , and P is the corresponding projector (which projects S onto V), then the operator PTP will be called the V-component of T .

If T is a PCO, then PTP is also a PCO.

The following theorem holds:

15.X. Let $\{P_n\}$ be a sequence of projectors converging strongly to the operator P (which is necessarily a projector). Let T be a linear compact operator. Then the sequence $\{P_n T P_n\}$ converges uniformly to PTP.

If T_1 and T_2 are linear hermitian operator we say that $T_1 > T_2$ if $T_1 - T_2$ is positive .

15.XI. If T_1 and T_2 are PCO and $T_1 < T_2$, then we have:

$$\mu_{1k} \leq \mu_{2k} \qquad (k = 1, 2, \cdots) ,$$

where $\{\mu_{1k}\}$ and $\{\mu_{2k}\}$ are, respectively, the sequences of eigenvalues of T_1 and T_2 .

This theorem is a consequence of the mini-max principle.

15.XII. Let T be a PCO with eigenvalues $\{\mu_k\}$, and let V

<u>be a subspace of</u> S . <u>If</u> $\{\nu_k\}$ <u>is the sequence of eigenvalues of the</u> V <u>-component of</u> T , <u>then</u> $\nu_k \leq \mu_k$ $(k = 1, 2, \ldots)$.

This theorem also follows from the mini-max principle.

Note that theorem 15.XII is <u>not</u> a particular case of 15.XI since, in general, it is not true that $T > PTP$.

We are now in a position to formulate the Rayleigh-Ritz method for the approximation of eigenvalues of a PCO T .

Let $\{v_k\}$ be a complete system of linearly independent vectors in the space S . Let us denote by V_n the n-dimensional subspace spanned by $v_1, \ldots, v_n$. We denote by P_n the projector of S onto V_n . The Rayleigh-Ritz method consists in considering the eigenvalues $\mu_1^{(n)}, \ldots, \mu_k^{(n)}, \ldots$ of $P_n T P_n$ as approximations of the eigenvalues $\mu_1, \ldots, \mu_k, \ldots$ of T . We shall prove that:

15.XIII i) <u>The eigenvalues of</u> $P_n T P_n$ <u>are the roots of the following determinantal equation:</u>

$$\det\{(Tv_i, v_j) - \mu(v_i, v_j)\} = 0 \qquad (i, j = 1, \ldots, n) \tag{15.4}$$

<u>plus the eigenvalue</u> $\mu = 0$;

ii) <u>The sequence</u> $\mu_k^{(n)}$ <u>does not decrease when</u> n <u>increases,</u> i.e. $\mu_k^{(n)} \leq \mu_k^{(n+1)}$ <u>for any</u> k ;

iii)
$$\lim_{n \to \infty} \mu_k^{(n)} = \mu_k$$

<u>uniformly with respect to</u> k .

It is immediately seen that the operator $P_n T P_n$ has the eigenvalue

$\mu = 0$ and that the corresponding eigenset (subspace of the solutions of the homogeneous equation $P_n T P_n u - \mu u = 0$ for $\mu = 0$) contains the subspace $S \ominus V_n$. The remaining eigenvalues and eigenvectors are obtained through the equations:

(15.5) $$P_n T P_n u - \mu P_n u = 0 \qquad , u = P_n u .$$

Due to (15.2), equation (15.5) can be written in the form:

(15.6) $$\alpha_{hk} \alpha_{ij} (u, v_i)(T v_j , v_h) - \mu \alpha_{hk} (u, v_h) = 0$$
$$(k = 1, \dots, n).$$

Since $\det \{\alpha_{hk}\} \neq 0$, setting $(u, v_i) = c_i$, we obtain:

(15.7) $$\alpha_{ij} (T v_j , v_h) c_i - \mu c_h = 0 \qquad (h = 1, \dots, n).$$

Equation (15.7) and (15.5) are equivalent.

Put $\gamma_j = \alpha_{ij} c_i$. Then $c_h = (v_j , v_h) \gamma_j$. Equation (15.7) is equivalent to the following:

(15.8) $$(T v_j , v_h) \gamma_j - \mu (v_j , v_h) \gamma_j = 0 .$$

It follows that μ is an eigenvalue for (15.5) if and only if is a root of equation (15.4). If $\tilde{\gamma}_1 , \dots , \tilde{\gamma}_n$ is an eigansolution for the algebraic system (15.8) corresponding to the eigenvalue $\tilde{\mu}$, then $u = \tilde{\gamma}_j v_j$ is an eigensolution for (15.5). The multiplicity of the eigenvalue $\tilde{\mu}$ of (15.5) is the nullity of the matrix $\{(Tv_i, v_j) - \tilde{\mu}(v_i, v_j)\}$. If $\tilde{\mu} \neq 0$, then this nullity is also the multiplicity of $\tilde{\mu}$ as an eigenvalue of $P_n T P_n$.

Statement ii) follows from theorem 15.XII. In fact, $P_n T P_n$ is the V_n-component of $P_{n+1} T P_{n+1}$ since $P_n P_{n+1} = P_{n+1} P_n = P_n$.

Due to the completeness of the system $\{v_h\}$, the projector P_n converges strongly to the identity operator. Therefore, from theorems 15.X and 15.IX, statement iii) follows.

Bibliography of Lecture 15

[1] N. ARONSZAJN - The Rayleigh-Ritz and the Weinstein Methods for approximation of Eigenvalues- (1. Operators in Hilbert Space) - Dept. of Math. Oklahoma Agricultural and Mechanical College, Stillwater Oklahoma - Tech. Rep. N° 1, 1949.

[2] G. FICHERA - see [1] of lecture 1.

[3] S.H. GOULD - Variational Methods in Eigenvalue, Problems - Univ. of Toronto Press, Toronto, 1957.

[4] F. RIESZ - B. Sz. NAGY - see [5] of lecture 2.

[5] A. WEINSTEIN - The intermediate Problems and the Maximum-Minimum Theory of Eigenvalues - Jour. of Math. and Mech. Indiana Univ. vol. 12, 1963.

L e c t u r e   16

The Weinstein - Aronszajn method.

The Raylegh-Ritz method provides a strong tool for the lower approximation of the eigenvalues of a PCO. Much more difficult is the problem consisting in the upper approximation of these eigenvalues, i. e. the construction of a sequence converging by decreasing to every given eigenvalue μ_K of T.

A method for solving this problem was proposed by A.Weinstein. His theory has been extended by Aronszajn and applied to several problems by different authors.

Weinstein's idea is the following. First he consider a PCO operator T_o, whose eigenvalues $\{\sigma_K\}$ and corresponding eigensolutions $\{w_K\}$ are known, and which is such that $\sigma_K > \mu_K$ for any K. Then he constructs a sequence T_n of PCO's such that:

i) If $\{\sigma_K^{(n)}\}$ are the eigenvalues of T_n, then $\sigma_K \geq \sigma_K^{(n)} \geq$ $\geq \sigma_K^{(n+1)} \geq \mu_K$;

ii) The eigenvalues of T_n can be computed in terms of the eigenvalues of T_o.

If $\{T_n\}$ converges uniformly to T, then $\sigma_K^{(n)}$ converges to μ_K (theor. 15.IX) by decreasing.(1)

---

(1) We use the name "decreasing sequence" for a sequence $\sigma^{(n)}$ such that $\sigma^{(n)} \geq \sigma^{(n+1)}$.

The original method of Weinstein, which was concerned with the eigenvalues problems arising in the theory of thin plates, considered as operator T a PCO which was the V -component of another PCO T_o, V being a given subspace of S. Then, denoting by P the projector which projects S onto $S \ominus V$ and by $\{v_k\}$ a complete system in $S \ominus V$, the "intermediate" operators T_n are given by $(I-P_n)T_o(I-P_n)$, where P_n is the projector on the subspace spanned by $v_1, \ldots, v_n$. The above stated condition i) for T_n is then a consequence of theor. 15.XII. Convergence follows from 15.X. Condition ii) will be proved later.

Aronszajn proposes to consider as T_o any operator such that $T_o > T$ (following in this respect a suggestion of H. Weyl), and then he constructs a sequence $\{T_n\}$ such that $T_o > T_n > T_{n+1} > T$ which satisfies condition ii).

In order to include in our treatment both the methods of Weinstein and of Aronszajn, it is convenient to give some general theorems. However we must first consider the resolvent operator for the following equation:

$$(16.1) \qquad Tu - \mu u = v \qquad (\mu \neq 0)$$

where T is a PCO. Let us distinguish two cases.

1) μ is not an eigenvalue for T. In this case equation (16.1) has one and only one solution given by:

$$(16.2) \qquad u \equiv R_\mu v = \frac{1}{\mu}\left[\sum_k \frac{\mu_k}{\mu_k - \mu}(v, u_k)u_k - v\right].$$

This result follows easily from (15.1).

2) μ is an eigenvalue for T. Then (16.1) has solutions if and only if $(v, u) = 0$, where u is any eigenvector corresponding to the eigenvalue μ. If this condition is satisfied, then every solution of (16.2) is given by

$$u = \tilde{R}_\mu v + a_1 u^{(1)} + \cdots + a_s u^{(s)} =$$

(16.3)

$$= \frac{1}{\mu} \left[\sum_k^{(\mu)} \frac{\mu_k}{\mu_k - \mu} (v, u_k) u_k - v \right] + \sum_{i=1}^{s} a_i u^{(i)},$$

where $\sum_k^{(\mu)}$ means that all the terms corresponding to indeces k for which $\mu_k = \mu$ were suppressed from the summation. The $u^{(1)}, \ldots, u^{(s)}$ are a complete system of eigenvectors in the eigenset corresponding to the eigenvalue μ. The a_i are arbitrary constants. This result is also easily obtained by using (15.1).

The following theorem constitutes the theoretical foundation of Weinstein's method.

16.I. Let T_0 and T be two PCO's. Let $\{T_n\}$ be a sequence of PCO's such that: i) $T_n = T_0 - D_n$ where D_n is a degenerate operator[2]; ii) $\{T_n\}$ converges uniformly to the operator T. Then any eigenvalue $\mu_k^{(n)}$ of T_n can be computed in terms of the eigenvalues and of the eigensolutions of T_0 and

$$\lim_{n \to \infty} \mu_k^{(n)} = \mu_k .$$

---

[2] We say that D is a degenerate operator if its range is finite dimensional.

The only thing we have to prove is that if T_0 is a PCO and D is an hermitian degenerate operator, then, given that the eigenvalues $\{\sigma_k\}$ of T_0 and the corresponding eigensolutions $\{w_k\}$ are known, it is possible to compute the eigenvalues of $T_0 - D$.

Let us suppose that the range of D is m-dimensional. Let $v_1, \dots, v_m$ be a basis in this range. It is easy to prove that there exists a non-singular $m \times m$ hermitian matrix $\{d_{ij}\}$ such that D admits the following representation: (3)

$$Du = d_{ij}(u, v_i)v_j .$$

We wish to find the non-zero eigenvalues of the following homogeneous equation:

$$T_0 u - \sigma u = d_{ij}(u, v_i)v_j . \tag{16.4}$$

Let us first suppose that $\sigma \neq 0$ is not an eigenvalue for T_0. Then, by using (16.2), we see that equation (16.4) is equivalent to the following equations:

$$u = d_{ij} c_i R_\sigma v_j \quad (16.5) ; \qquad c_h = (u, v_h). \quad (16.6)$$

---

(3) Let $Du = c_j(u)v_j$. We have $(Du, v_i) = c_j(u)(v_j, v_i)$. Then $c_j(u) = \alpha_{ij}(Du, v_i) = \alpha_{ij}(u, Dv_i)$. Set $Dv_i = \gamma_{is} v_s$. We have $c_j(u) = \alpha_{ij}\bar{\gamma}_{is}(u, v_s)$. It follows that $Du = d_{ij}(u, v_i)v_j$, where $d_{ij} = \bar{\alpha}_{hj}\bar{\gamma}_{hi}$. Since D is hermitian, so is $\{d_{ij}\}$.

By inserting (16.5) into (16.6) we get

(16.7) $$(R_\sigma v_j, v_h)\, d_{ij}\, c_i - c_h = 0 .$$

It is obvious that σ (not belonging to the spectrum of T_0) is an eigenvalue for $T_0 - D$ if and only if the algebraic system (16.7) has non trivial solutions. If $\{a_{jh}\}$ is the inverse matrix of $\{d_{ij}\}$, then we can write (16.7) in the equivalent form:

$$(R_\sigma v_j, v_h)\, \gamma_j - a_{jh}\, \gamma_j = 0 .$$

It follows that σ (not belonging to the spectrum of T_0) is an eigenvalue for $T_0 - D$ if and only if it is a solution of the equation:

(16.8) $$W(\sigma) \equiv \det\left\{(R_\sigma v_j, v_h) - a_{jh}\right\} = 0 .$$

The determinant $W(\sigma)$ is the Weinstein determinant, which was introduced by that author in connection with the above mentioned particular case.

Let us now suppose that $\sigma \neq 0$ is an eigenvalue for T_0. We denote by $u^{(1)}, \dots, u^{(s)}$ a complete system of eigenvectors of T_0 corresponding to σ. If u satisfies (16.4), the constants c_h defined by (16.6) must satisfy the equations:

(16.9) $$d_{ij}\,(v_j, u^{(p)})\, c_i = 0 \qquad p = 1, \dots, s .$$

Moreover u must be given by [see (16.3)] :

$$u = d_{ij} c_i \tilde{R}_\sigma v_j + \sum_{p=1}^{s} a_p u^{(p)}. \tag{16.10}$$

Inserting (16.10) into (16.6) we get:

$$(\tilde{R}_\sigma v_j, v_h) d_{ij} c_i - c_h + \sum_{p=1}^{s} (u^{(p)}, v_h) a_p = 0. \tag{16.11}$$

It follows that if the eigenvalue σ of T_o is an eigenvalue for (16.4), then the homogeneous system (16.11), (16.9) has non trivial solutions $c_1, \dots, c_m, a_1, \dots, a_s$.

Conversely, if (16.11), (16.9) has non trivial solutions, then - as is easily seen - the eigenvalue σ of T_o is an eigenvalue for $T_o - D$. It follows that an eigenvalue σ of T_o is an eigenvalue of $T_o - D$ if and only if it is a solution of the equation:

$$\det \left\{ \begin{array}{c|c} (\tilde{R}_\sigma v_j, v_h) - a_{jh} & (u^{(p)}, v_h) \\ \hline (v_j, u^{(p)}) & 0 \end{array} \right\} = 0. \tag{16.12}$$

The meaning of the $(m+s) \times (m+s)$ matrix between the brackets { } is obvious.

In conclusion, in order to compute the eigenvalues of $T_0 - D$, we have first to solve equation (16.8). Roots of this equation which are not in the spectrum of T_0 are eigenvalues for $T_0 - D$. Then we solve equation (16.12). Roots of this equation which are eigenvalues for T_0, are also eigenvalues for $T_0 - D$. It is easily seen that the multiplicity of the eigenvalue of $T_0 - D$ is given by the nullity of the matrix (16.8) or (16.12) respectively.

From the theorem we have just proved, it follows that condition ii) for the intermediate operator T_n is satisfied if $T_0 - T_n$ is degenerate. It is immediate that this condition is satisfied for the intermediate operators of the original Weinstein method. Since:

$$T_n = (I - P_n) T_0 (I - P_n) = T_0 - (P_n T_0 + T_0 P_n - P_n T_0 P_n),$$

the operator $D_n = P_n T_0 + T_0 P_n - P_n T_0 P_n$ is degenerate.

The ability to apply theorem 16.I rests on the possibility of approximating the compact operator $T_0 - T$ by a sequence of degenerate operators. It is known that this can be done in infinitely many ways in a Hilbert space. If we further require that the approximating sequence be such that for T_n the condition i) is satisfied, then we can apply Weinstein's method.

There are several methods for constructing the intermediate operators T_n which have been proposed by different authors. Later we shall give a general procedure which is particularly suitable for eigenvalue problems connected with BVP's for elliptic systems.

We wish first to show how to overcome one difficulty which arises in applying Weinstein's method.

We have seen that the eigenvalues of the intermediate operators can be found by solving equations (16.8) and (16.12). This problem appears to be a very difficult one, in the general case, since it consists in determining the zeros of meromorphic functions.

Two procedures have been proposed in order to overcome this difficulty. The first of these was developed by Weinstein himself and, later, by Bazley [3]. It is called the <u>method of special choices</u>. The second is due to Weinberger[7] and is known as the <u>method of truncation</u>.

The special choices consist simply in constructing the intermediate operators $T_0 - D$ (we omit the index n from D) in such a way that the range V of D is spanned by vectors $v_1, \dots, v_m$ such that for any j, the scalar product (w_k, v_j) is different from zero only for a finite set of values of the index k.(4) This condition is, for instance, satisfied when V has as a basis m eigenvectors of the operator T_0 (Bazley), or when V has as a basis m vectors belonging to a system which together with the system of eigenvectors $\{w_k\}$ of T_0 forms a pair of biorthogonal systems.

If the above condition is satisfied, then, from (16.2) and (16.3) it follows readily that solving equations (16.8) and (16.12) consists in finding zeros of polynomials.

The method of truncation consists in replacing the operator T_0 by the <u>truncated</u> operator

$$T_0^{(m)} u = \sum_{k=1}^{m} \sigma_k (u, w_k) w_k + \sigma_{m+1} \left[u - \sum_{k=1}^{m} (u, w_k) w_k \right].$$

---

(4) $\{w_k\}$ is the system of eigenvectors of T_0.

We denote - as usual - by $\{\sigma_\kappa\}$ the eigenvalues of T_0. As intermediate operators we take $T_n^{(m)} = T_0^{(m)} - D_n$. The solution of the equation $T_0^{(m)} u - \sigma u = v$ is given by:

$$u = R_\sigma^{(m)} v = \sum_{\kappa=1}^{m} \frac{\sigma_{m+1} - \sigma_\kappa}{(\sigma_{m+1} - \sigma)(\sigma_\kappa - \sigma)} (v, w_\kappa) w_\kappa + \frac{v}{\sigma_{m+1} - \sigma}$$

if σ does not belong to the spectrum of $T_0^{(m)}$.

In the case $\sigma = \sigma_\kappa$ $(\kappa = 1, 2, .., m)$ the most general solution of the above equation is:

$$u = \tilde{R}_\sigma^{(m)} v = \sum_{\kappa=1}^{m}{}^{(\sigma)} \frac{\sigma_{m+1} - \sigma_\kappa}{(\sigma_{m+1} - \sigma)(\sigma_\kappa - \sigma)} (v, w_\kappa) w_\kappa +$$

$$+ \frac{v}{\sigma_{m+1} - \sigma} + \sum_{i=1}^{s} a_i u^{(i)},$$

where $\sum^{(\sigma)}$ has the usual meaning and $\{u^{(i)}\}$ $(i = 1, .., s)$ is a complete system of eigenvectors of T_0 corresponding to the eigenvalue σ.

By using the same arguments as in the proof of theorem 16.I, it is easy to see that the eigenvalues of $T_0^{(m)} - D_n$ satisfying the condition $\sigma \neq \sigma_{m+1}$, are obtained by solving algebraic equations.

If $\sigma_1^{(m,n)} \geq \sigma_2^{(m,n)} \geq .. \geq \sigma_{p_{m,n}}^{(m,n)}$ are the eigenvalues of $T_0^{(m)} - D_m$ greater than σ_{m+1}, it is easy to prove that if condition ii) of theor. 16.I is satisfied,

$$\lim_{\substack{m \to \infty \\ n \to \infty}} \sigma_\kappa^{(m,n)} = \mu_\kappa .$$

Bibliography of Lecture 16

[1] N.ARONSZAJN - see [1] of lecture 15.

[2] N.ARONSZAJN - Approximation methods for eigenvalues of completely continuous symmetric operators - Proc. Symp. Spectral Theory and Diff. Problems - Stillwater, Okla. 1951.

[3] N.W.BAZLEY - Lower Bounds for Eigenvalues - Journ. Math.& Mech. vol.10, 1961.

[4] N.W.BAZLEY-D.W.FOX - Truncations in the Method of Intermediate Problems for Lower Bounds to Eigenvalues - Journ. of Research of the N.B.S. - Math. and Math.Phy. vol. 56 B, 1961.

[5] G.FICHERA - see [1] of lecture 1.

[6] S.H.GOULD - see [3] of lecture 15.

[7] H.F.WEINBERGER - A theory of lower bounds for eigenvalues - Tech. Note BN-183 (Inst. Fluid Dynam. & Appl. Math. Univ. of Maryland),1959.

[8] A.WEINSTEIN - Études des spectres des équations aux dérivées partielles - Mémorial des Sci. Mathem. No.88, 1937.

Lecture 17

Construction of the intermediate operators.

The Weinstein-Aronszajn method described in the previous lecture requires the construction of "intermediate operators" T_n between the "base operator" T_0 and the given operator T. The intermediate operators are required to satisfy the following conditions:

i) $$\sigma_k \geq \sigma_k^{(n)} \geq \sigma_k^{(n+1)} \geq \mu_k$$

($\{\sigma_k\} \equiv$ eigenvalues of T_0, $\{\sigma_k^{(n)}\} \equiv$ eigenvalues of T_n, $\{\mu_k\} \equiv$ eigenvalues of T).

ii) $T_0 - T_n$ is degenerate.

iii) T_n converges uniformly to T.

Condition i) is satisfied if $T_0 > T_n > T_{n+1} > T$.

The original Weinstein method, where (see previous lecture) $T = (I-P)T_0(I-P)$ and $T_n = (I-P_n)T_0(I-P_n)$ does not satisfy this monotonicity condition, but does satisfy condition i). However, we shall prove that, even in Weinstein's case, one can always reduce to the above monotonicity situation.

Set $Q = I-P$ and $Q_n = I-P_n$. Let us first prove that:

17.I. <u>Any non-zero eigenvalue of the operator</u> $Q T_0 Q$ <u>is an eigenvalue for the following problem:</u>

$$T_0 u - P T_0 u = \mu u, \tag{17.1}$$

<u>and, conversely, any non-zero eigenvalue for (17.1) is an eigenvalue for</u> $Q T_0 Q$.

Let $\mu \neq 0$ and u be an eigenvalue and eigenvector for $Q T_0 Q$. From:

$$Q T_0 Q u = \mu Q u + \mu P u \tag{17.2}$$

it follows that $u = Q u$. Therefore, we can write (17.2) as follows: $Q T_0 u = \mu u$, i.e. (17.1). Conversely, if $\mu \neq 0$ and $u \neq 0$ satisfy (17.1), then $u = \frac{1}{\mu} Q T_0 u$ and hence, $u = Q u$. Thus $Q T_0 Q u = \mu u$. Of course, the theorem still holds if we replace P by P_n.

Let us now suppose that T_0 is strictly positive. Then we have:

17.II. <u>Any eigenvalue for the problem (17.1) is an eigenvalue for:</u>

$$T_0 v - T_0^{\frac{1}{2}} P T_0^{\frac{1}{2}} v = \mu v \qquad \left(T_0^{\frac{1}{2}} > 0\right). \tag{17.3}$$

<u>Conversely, any non-zero eigenvalue for (17.3) is an eigenvalue for (17.1).</u>

If we let $v = T_0^{\frac{1}{2}} u$, then equation (17.1) becomes:

$$T_0^{\frac{1}{2}} v - P T_0^{\frac{1}{2}} v = \mu u.$$

Operating on both sides by $T_0^{\frac{1}{2}}$, we get:

$$T_0 v - T_0^{\frac{1}{2}} P T_0^{\frac{1}{2}} v = \mu v.$$

Conversely, let us assume that $\mu \neq 0$ and v satisfy (17.3). From (17.3) we deduce that there exists a vector u such that $v = T_0^{\frac{1}{2}} u$. We can write (17.3) as follows:

$$T_0^{\frac{1}{2}} [T_0 u - P T_0 u - \mu u] = 0.$$

Since $T_0^{\frac{1}{2}}$ is strictly positive, (17.1) must be satisfied.

From the above theorems it follows that in Weinstein's original method we can solve as intermediate problem any of the following:

a) $(I - P_n) T_0 (I - P_n) u - \mu u = 0$

b) $T_0 u - P_n T_0 u - \mu u = 0$

c) $T_0 u - T_0^{\frac{1}{2}} P_n T_0^{\frac{1}{2}} u - \mu u = 0.$

Actually, in his work, Weinstein considered problems b) as intermediate problems.

If we consider problem c), assume as T_n the operator $T_0 - T_0^{\frac{1}{2}} P_n T_0^{\frac{1}{2}}$, and replace T by $T_0 - T_0^{\frac{1}{2}} P T_0^{\frac{1}{2}}$ (this is feasible because of theorems 17.I and 17.II), then the above mentioned monotonicity condition is clearly satisfied.

In considering a general method for constructing the intermediate operators, we shall assume that the base operator T_0 is greater than T and shall replace condition i) by the following:

i') $$T_0 > T_m > T_{m+1} > T.$$

Our construction includes as particular cases the procedures given by Weinstein and Aronszajn and some of the methods considered by Bazley and Bazley & Fox. [1]

Set $L = T_0 - T$ and assume that $L = \sum_{i=1}^{q} L_i$, where each L_i is a PCO. Let us associate with L_i a linear operator M_i which is supposed to be strictly positive. We denote by S_i the Hilbert space which is obtained by the completion of S with respect to the following new scalar product:

$$(u, v)_i = (M_i u, v).$$

The operator M_i can be extended in the space S_i in such a way that its extension is strictly positive in S_i and its range still belongs to S. In fact, for $u \in S$ and $v \in S_i$, let us consider the scalar product $(u,v)_i$. We have $|(u,v)_i| \leq c_i |u| |v|_i$, where $|\ |$ ($|\ |_i$) denotes the length of a vector in the space S (S_i). Since $(u,v)_i$ is a linear bounded functional in the space S, there exists $\tilde{M}_i$ such that $(u,v)_i = (u, \tilde{M}_i v)$ and $\tilde{M}_i v \in S$. For $v \in S$ we have obviously $\tilde{M}_i v = M_i v$. On the other hand $(\tilde{M}_i v, v)_i = (\tilde{M}_i v, \tilde{M}_i v)$ and $\tilde{M}_i v = 0$ implies $v = 0$. It follows that $\tilde{M}_i$ is the desired extension of M_i. We shall use the same symbol M_i to denote the extension $\tilde{M}_i$ of the operator under consideration.

Let us now introduce a new Hilbert space H_i. The scalar product in H_i will be denoted by $[\ ,\ ]_i$. By R_i we denote a compact

---

(1) It must be observed that these authors consider eigenvalue problems for more general operators than PCO's. However the main ideas do not differ substantially from the PCO case.

linear operator with domain S_i and range in the space H_i. Since R_i is bounded, there exists its adjoint operator R_i^*, that is to say, a linear bounded transformation of H_i into S_i such that for any $u \in S_i$ and any $v \in H_i$:

$$[R_i u, v]_i = (u, R_i^* v)_i .$$

We shall suppose that L_i admits the following decomposition:

(17.4) $$L_i = M_i R_i^* P^{(i)} R_i \quad ,$$

where $P^{(i)}$ is any given projector of the space H_i onto one of its subspaces V_i. The decomposition (17.4) is admissible since L_i, if represented by (17.4), is an operator which maps S into S as follows from the diagram:

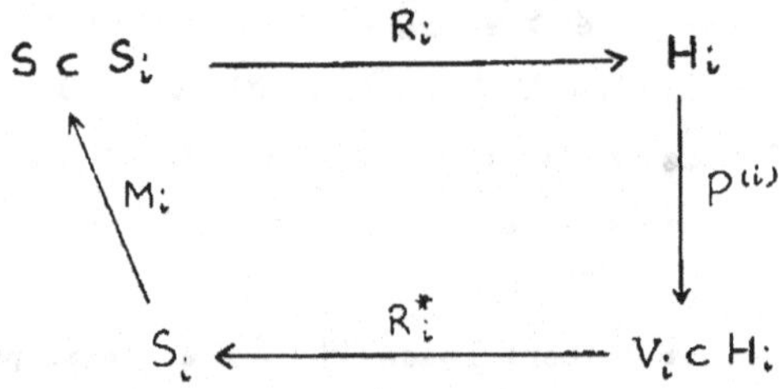

On the other hand we have:

$$(M_i R_i^* P^{(i)} R_i u, v) = (R_i^* P^{(i)} R_i u, v)_i =$$

$$= [P^{(i)} R_i u , R_i v]_i = [P^{(i)} R_i u, P^{(i)} R_i v]_i =$$

$$= (u, R_i^* P^{(i)} R_i v)_i = (u, M_i R_i^* P^{(i)} R_i v).$$

From these equations it follows that L_i is a PCO.

Let us now suppose that V_i is separable and denote by $\{\omega_i^{(s)}\}$ a complete system of linearly independent vectors in the subspace V_i of H_i. Let $P_n^{(i)}$ be the projector of H_i onto the n-dimensional variety spanned by $\omega_i^{(1)}, \dots, \omega_i^{(n)}$.

17.III. The sequence:

$$T_n = T_0 - \sum_{i=1}^{q} M_i R_i^* P_n^{(i)} R_i$$

is a sequence of intermediate operators, i.e. conditions i),ii),iii) are satisfied.

We have for any $u \in S$,

$$(T_0 u, u) \geq (T_0 u, u) - \sum_{i=1}^{q} [P_n^{(i)} R_i u, P_n^{(i)} R_i u]_i = (T_n u, u) \geq$$

$$\geq (T_{n+1} u, u) \geq (T_0 u, u) - \sum_{i=1}^{q} [P^{(i)} R_i u, P^{(i)} R_i u]_i = (Tu, u).$$

This proves that condition i') is satisfied.

The operator:

$$T_0 - T_n = \sum_{i=1}^{q} M_i R_i^* P_n^{(i)} R_i$$

is degenerate since $P_n^{(i)}$ projects onto a finite dimensional subspace.

In order to prove iii), it suffices to show that:

$$\lim_{n \to \infty} \| P^{(i)} R_i - P_n^{(i)} R_i \| = 0. \tag{17.5}$$

The unit sphere $|u|_i = 1$ of S_i is mapped by $P^{(i)} R_i$ into a compact subset of V_i. Let $\{v_i^{(k)}\}$ be an orthormal complete set in the space V_i such that $P_n^{(i)} v = \sum_{k=1}^{n} (v, v_i^{(k)}) v_i^{(k)}$. Then $\| P^{(i)} R_i u - P_n^{(i)} R_i u \|_i^2 = \sum_{k>n} | [R_i u, v^{(k)}] |^2$ tends to zero, for $n \to \infty$, uniformly with respect to u if $|u|_i = 1$ (2). From this, (17.5) follows, and the proof of the theorem is complete.(3)

If we assume that $T_0 - T$ is strictly positive, $q = 1$, $S_1 = H_1$, $M_1 = L_1$, $R_1 = P^{(1)} = I$, then we have:

$$T_n = T_0 - L_1 P_n$$

where P_n projects S_1 into the space spanned by $\omega_1, \dots, \omega_n$ and $\{\omega_k\}$ is any complete system in the space S_1. These T_n are the intermediate operators constructed by Aronszajn [1].

If $q = 1$ and $L = T_0^{\frac{1}{2}} P T_0^{\frac{1}{2}}$, where P is any projector of the space S, then the above general procedures give us as a particular case the intermediate problems c) which are to be equivalent to Weinstein's intermediate problems.

---

(2) See footnote (1) of pag.20.
We have denoted by $\| \ \ \|_i$ the norm in the space H_i.

(3) The proof of iii) can be carried out by a similar procedure if we replace the hypotheses of compactness of R_i by compactness of M_i.

If we assume $q = 1$, $M_1 = I$ and $S = S_1 = H_1$, $R_1 = L^{\frac{1}{2}}$ $(L^{\frac{1}{2}} > 0)$, $P^{(1)} = I$, we have the following kind of intermediate operators: $T_n = T_0 - L^{\frac{1}{2}} P_n L^{\frac{1}{2}}$, where P_n is the projection onto the subspace spanned by the first n vectors of any complete system in the space S .

It is now evident how to construct as many examples as we wish, starting from the general procedure.

Bibliography of Lecture 17

[1] N.ARONSZAJN - see [2] of lecture 16.

[2] N.W.BAZLEY-D.W.FOX - Lower Bounds to Eigenvalues using operator Decompositions of the form B^*B - Arch. for Rat. Mech. and Anal. vol 10, 1962.

[3] N.W.BAZLEY-D.W.FOX - Improvement of Bounds to Eigenvalues of Operators of the form T^*T - The Johns Hopkins Univ. Appl. Phy. Lab. (Report) 1964.

[4] N.W.BAZLEY-D.W.FOX - Comparison Operators for Lower Bounds to Eigenvalues - Battelle Centre de recherche de Geneva- (Report) 1963.

[5] N.W. BAZLEY-D.W.FOX - Methods for Lower Bounds to Frequencies of Continuous Elastic Systems - The Johns Hopkins Univ. Appl. Phy. Lab. (Report), 1964.

[6] G.FICHERA - Sul calcolo degli autovalori - Atti del Convegno sulle applicazioni dell'Analisi alla Fisica Matem. - Cagliari-Sassari 1964.

[7] G.FICHERA - Approximations and Estimates for Eigenvalues of BVP. - Proc. of the Symposium on the Numerical Solution of PDE - Univ. of Maryland (to appear).

[8] S.T.KURODA - On a Generalization of the Weinstein-Aronszajn Formula and the Infinite Determinant - Rep. from Sci. Papers of the College of Gen. Education, Univ. of Tokyo - vol. 11 - N° 1 , 1961.

Lecture 18

Orthogonal invariants of positive compact operators.

The method developed in lectures 16 and 17 must be considered, after the essential contributions to it by Aronszajn, Weinberger and Bazley, as a very efficient tool for the upper approximation of eigenvalues of a PCO. However a serious limitation to its applicability is the requirement that a "base operator" T_0 must be known together with the entire set of its eigenvalues and eigenvectors.

In order to clarify this point, let us suppose that T is an integral operator in the Hilbert space $\mathcal{L}^2(0,1)$ given by :

$$Tu = \int_0^1 K(x,y)\, u(y)\, dy.$$

The kernel $K(x,y)$ is supposed to belong to $\mathcal{L}^2[(0,1)\times(0,1)]$, to be hermitian, i.e. $K(x,y) = \overline{K(y,x)}$ and to be of "positive type", i.e.:

$$\int_0^1\int_0^1 K(x,y)\,\overline{u(x)}\,u(y)\,dx\,dy \geq 0$$

for every $u \in \mathcal{L}^2(0,1)$. If we wish to apply the above-mentioned method for the upper approximation of the eigenvalues of the kernel $K(x,y)$, we have to know a hermitian kernel $K_0(x,y)$ such that any of its eigenvalues is greater than the corresponding eigenvalue of $K(x,y)$. Moreover we must know every eigenvalue and every eigenvector of the kernel $K_0(x,y)$. In general, we do not know to construct the kernel $K_0(x,y)$.

We wish now to develop a different method, which must be considered as an alternative to the Weinstein-Aronszajn one. Its applicability requires a further condition on the operator T (T must belong to one of the classes $\mathcal{T}^n$, which will be defined later), however the new method will not require the assumption of the existence of a base operator T_0. On the other hand, while the first method can be applied - under special conditions - to non-compact operators, the extension of the second one to these more general cases it not yet known.

Let us consider a PCO T in the space S (which is a separable complex infinite dimensional Hilbert space). For the sake of simplicity, we shall suppose from now on that T is strictly positive and that PCO stands for a strictly positive compact operator. The reader will notice by himself the slight modifications of the following results which must be made in order to include positive operators which are not strictly positive.

We shall denote by $G^{(n)}(v_1,\dots,v_s)$ the Gramian determinant of the s vectors $v_1, \dots, v_s$ with respect to the scalar product

$$(T^n u, v)$$

where n is a positive integer. In other words, we set, by definition,

$$G^{(n)}(v_1,\dots,v_s) = \begin{vmatrix} (T^n v_1, v_1) \dots\dots (T^n v_1, v_s) \\ \dots\dots\dots\dots \\ \dots\dots\dots\dots \\ (T^n v_s, v_1) \dots\dots (T^n v_s, v_s) \end{vmatrix}.$$

Let $\{v_h\}$ $(h = 1,2,\dots)$ be a complete orthonormal system in the space We put:

(18.1) $$\mathfrak{J}_0^n(T) = 1$$

and for any positive integer s,

(18.2) $$\mathfrak{J}_s^n(T) = \frac{1}{s!} \sum_{k_1,\dots,k_s} G^{(n)}(v_{k_1},\dots,v_{k_s}).$$

The summation is extended to any set of s positive integers $k_1,\dots,k_s$. Since the terms of the multiple series are non-negative, it does not matter how the summation is carried out. Of course the value of $\mathfrak{J}_s^n(T)$ could be finite or infinite. It is evident that:

$$\mathfrak{J}_s^n(T^m) = \mathfrak{J}_s^{n+m}(T).$$

18.I <u>The value of $\mathfrak{J}_s^n(T)$ does not depend on the orthonormal system $\{v_k\}$, i.e. $\mathfrak{J}_s^n(T)$ is an orthonormal invariant of the operator</u> T.

In order to prove this important theorem we need first the following lemma.

18.II. <u>If</u> $G(u_1,\ldots,u_n)$ <u>is the Gramian determinant of</u> n <u>vectors in a Hilbert space</u> S <u>, and</u> $0 < k < n$ <u>, then the following inequality holds:</u>

$$G(u_1,\ldots,u_n) \leq G(u_1,\ldots,u_k)\,G(u_{k+1},\ldots,u_n).$$

The proof is trivial if $u_1,\ldots,u_n$ are linearly dependent vectors. Let us suppose that $G(u_1,\ldots,u_n) > 0$ and denote by S_n the subspace spanned by $u_1,\ldots,u_n$. Let $x_k^1,\ldots,x_k^n$ be the coordinates of u_k with respect to an orthonormal basis in S_n. We have:

$$G(u_1,\ldots,u_n) = \begin{vmatrix} x_1^1 & \cdots & x_1^n \\ \cdot & \cdots & \cdot \\ \cdot & \cdots & \cdot \\ x_n^1 & \cdots & x_n^n \end{vmatrix}^2 =$$

$$= \left[\sum (-1)^\sigma X_{1\ldots\ldots k}^{s_1\ldots s_k} X_{k+1\ldots\ldots n}^{s_{k+1}\ldots\ldots s_n} \right]^2,$$

where $\sigma = 1+\cdots+k+s_1+\cdots+s_k$ and $X_{j_1\ldots j_p}^{i_1\ldots i_p}$ denotes the determinant:

$$\begin{vmatrix} x_{j_1}^{i_1} & \cdots & x_{j_1}^{i_p} \\ \cdot & \cdots & \cdot \\ \cdot & \cdots & \cdot \\ x_{j_p}^{i_1} & \cdots & x_{j_p}^{i_p} \end{vmatrix}.$$

The summation is extended to any subdeterminant contained in the first k rows of the matrix $\{ x_j^i \}$ $(i, j = 1, \dots, n)$ (Laplace development of a determinant). It follows that:

$$G(u_1, \dots, u_n) \leq \sum \left(X_{1 \dots\dots k}^{s_1 \dots s_k} \right)^2 \sum \left(X_{k+1 \dots\dots n}^{s_{k+1} \dots s_n} \right)^2 =$$

$$= G(u_1, \dots, u_k)\, G(u_{k+1}, \dots, u_n).$$

We go now to the proof of theorem 18.I. Let us consider the spectral decomposition of the operator T^n:

$$T^n u = \sum_{h=1}^{\infty} \mu_h^n (u, u_h) u_h .$$

We have $(s > 0)$:

$$s!\, \mathcal{J}_s^n(T) = \lim_{m \to \infty} \sum_{k_1, \dots, k_s}^{1, \dots, m} \lim_{q \to \infty} \begin{vmatrix} \sum\limits_{h_1}^{1,q} \mu_{h_1}^n (v_{k_1}, u_{h_1}) \overline{(v_{k_1}, u_{h_1})} & \dots & \sum\limits_{h_s}^{1,q} \mu_{h_s}^n (v_{k_1}, u_{h_s}) \overline{(v_{k_s}, u_{h_s})} \\ \vdots & & \vdots \\ \sum\limits_{h_1}^{1,q} \mu_{h_1}^n (v_{k_s}, u_{h_1}) \overline{(v_{k_1}, u_{h_1})} & \dots & \sum\limits_{h_s}^{1,q} \mu_{h_s}^n (v_{k_s}, u_{h_s}) \overline{(v_{k_s}, u_{h_s})} \end{vmatrix}$$

Let us denote by $P_m w$ the projection of w on the variety spanned by $v_1, \dots, v_m$; $G(w_1, \dots, w_s)$ is the Gramian determinant of $w_1, \dots, w_s$ with respect to the scalar product (u,v). We have:

$$s! \, \mathcal{J}_s^n(T) = \lim_{m\to\infty} \lim_{q\to\infty} \sum_{h_1,\dots,h_s}^{1,\dots,q} \mu_{h_1}^n \cdots \mu_{h_s}^n \, G\,(P_m u_{h_1}, \dots, P_m u_{h_s}).$$

Let us now suppose that the multiple series $\sum_{h_1,\dots,h_s} \mu_{h_1}^n \cdots \mu_{h_s}^n$, where the summation is extended to any set $h_1, \dots, h_s$ of distinct indices is convergent. Let σ denote its sum and assume q_ε is such that:

$$\sigma - \sum_{h_1,\dots,h_s}^{1,\dots,q_\varepsilon} \mu_{h_1}^n \cdots \mu_{h_s}^n < \varepsilon,$$

where ε is a positive real number given arbitrarily.

Since, by lemma 18.II, we have:

$$G\,(P_m u_{h_1},\dots,P_m u_{h_s}) = |P_m u_{h_1}|^2 \dots |P_m u_{h_s}|^2,$$

it follows that:

$$\sum_{h_1,\dots,h_s} \mu^n_{h_1}\cdots\mu^n_{h_s} - \sum_{h_1,\dots,h_s} \mu^n_{h_1}\cdots\mu^n_{h_s}\, G(P_m u_{h_1},\dots,P_m u_{h_s}) \le$$

$$\le \sum_{h_1,\dots,h_s}^{1,\dots,q_\varepsilon} \mu^n_{h_1}\cdots\mu^n_{h_s}\left[1 - G(P_m u_{h_1},\dots,P_m u_{h_s})\right] + 2\varepsilon .$$

Thus,

(18.3) $$\mathcal{J}^n_s(T) = \frac{1}{s!}\sum_{h_1,\dots,h_s} \mu^n_{h_1}\cdots\mu^n_{h_s} = \sum_{h_1<\dots<h_s} \mu^n_{h_1}\cdots\mu^n_{h_s} .$$

Suppose the right-hand series in (18.3) is divergent. Given $H > 0$, let q_H be such that :

$$\sum_{h_1,\dots,h_s}^{1,\dots,q_H} \mu^n_{h_1}\cdots\mu^n_{h_s} > H .$$

Since:

$$\lim_{m\to\infty} \sum_{h_1,\dots,h_s} \mu^n_{h_1}\cdots\mu^n_{h_s}\, G(P_m u_{h_1},\dots,P_m u_{h_s}) \ge$$

$$\geq \lim_{m \to \infty} \sum_{h_1, \dots, h_s}^{1, \dots, q_H} \mu_{h_1}^n \cdots \mu_{h_s}^n G(P_m u_{h_1}, \dots, P_m u_{h_s}) > H ,$$

it follows that $\mathfrak{I}_s^n(T) = +\infty$. This means that (18.3) also holds in this case.

The index s will be called the <u>order</u> of the orthogonal invariant $\mathfrak{I}_s^n(T)$ and the index n the <u>degree</u> of this invariant.

18 .III. <u>We have</u> $\mathfrak{I}_s^n(T) < +\infty$ <u>if and only if</u> $\mathfrak{I}_1^n(T) < +\infty$.

The proof is a consequence of the following inequalities:

$$\mathfrak{I}_s^n(T) \leq \frac{1}{s!} \left[\mathfrak{I}_1^n(T) \right]^s . \tag{18.4}$$

$$\mathfrak{I}_1^n(T) \leq s! \, (\mu_1 \cdots \mu_{s-1})^{-n} \, \mathfrak{I}_s^n(T) + \sum_{k=1}^{s-1} \mu_k^n . \tag{18.5}$$

Since (lemma 18 .II):

$$G^{(n)}(v_{k_1}, \dots, v_{k_s}) \leq G^{(n)}(v_{k_1}) \cdots G^{(n)}(v_{k_s}) ,$$

(18.4) follows from (18.2). In order to prove (18.5) we observe that:

$$\mu_1^n \cdots \mu_{s-1}^n (\mu_s^n + \mu_{s+1}^n + \cdots) \leq s! \, \mathfrak{I}_s^n(T).$$

From this inequality (18.5) follows readily.

<u>A PCO is said to belong to the class</u> $\mathcal{T}^n$ <u>if</u> $\mathfrak{I}_1^n(T) < +\infty$.

It is evident that $\mathcal{T}^m \subset \mathcal{T}^n$ if $m < n$.

18.IV. <u>The sequence of positive numbers</u> $\{ \mathfrak{I}_s^n(T) \}$ $(s=1,2,\dots)$ <u>is a complete system of invariants with respect to the unitary equivalence of two PCO's of the class</u> $\mathcal{C}^n$.

We must prove that if T and R are two operators of $\mathcal{C}^n$ such that:

$$\mathfrak{I}_s^n(T) = \mathfrak{I}_s^n(R) \qquad (s=1,2,\dots)$$

then a unitary operator U of the space S exists such that:

$$T = U^{-1} R U \;; \tag{18.6}$$

i.e. the two operators are unitary equivalent.

Let us denote - as usual - by $\{\mu_k\}$ the sequence of eigenvalues of T (each repeated as many times as its multiplicity). The infinite product:

$$\prod_{k=1}^{\infty} (1-\mu_k^n \lambda)$$

converges uniformly on any compact set of the complex λ-plane and defines an entire function $\Delta(\lambda)$ of the complex variable λ. Let us define:

$$\Delta_m(\lambda) = \prod_{k=1}^{m} (1-\mu_k^n \lambda).$$

Let $\{u_k\}$ be a complete orthonormal set of eigenvectors of the operator T, with $Tu_k = \mu_k u_k$. Denote by P_m the projector which projects S onto the m-dimensional manifold spanned by $u_1, \dots, u_m$. We have:

$$\Delta_m(\lambda) = \sum_{s=0}^{m} (-1)^s \mathfrak{I}_s^n (P_m T) \lambda^s .$$

For any m, let us consider the power series:

$$\sum_{s=0}^{\infty} (-1)^s \mathfrak{I}_s^n (P_m T) \lambda^s . \quad ^{(1)}$$

It converges in the entire λ-plane uniformly with respect to m. This follows from the inequalities:

$$\left| \sum_{s=q+1}^{q+h} (-1)^s \mathfrak{I}_s^n (P_m T) \lambda^s \right| \leq \sum_{s=q+1}^{q+h} \mathfrak{I}_s^n (P_m T) |\lambda|^s \leq$$

$$\leq \sum_{s=q+1}^{q+h} \mathfrak{I}_s^n (T) |\lambda|^s \leq \sum_{s=q+1}^{q+h} \frac{1}{s!} \left[\mathfrak{I}_1^n (T) \right]^s |\lambda|^s .$$

On the other hand, for any given q we have:

$$\lim_{m \to \infty} \sum_{s=0}^{q} (-1)^s \mathfrak{I}_s^n (P_m T) \lambda^s = \sum_{s=0}^{q} (-1)^s \mathfrak{I}_s^n (T) \lambda^s .$$

Thus, for any complex λ,

$$\lim_{m \to \infty} \sum_{s=0}^{\infty} (-1)^s \mathfrak{I}_s^n (P_m T) \lambda^s = \sum_{s=0}^{\infty} (-1)^s \mathfrak{I}_s^n (T) \lambda^s .$$

---

(1) The operator $P_m T$ is considered as a strictly positive operator in the m-dimensional space spanned by $u_1, \dots, u_m$.

Set:

$$\tilde{\Delta}(\lambda) = \sum_{s=0}^{\infty} (-1)^s \mathcal{J}_s^n(T)\lambda^s, \quad \tilde{\Delta}^{(m)}(\lambda) = \sum_{s=0}^{\infty} (-1)^s \mathcal{J}_s^n(P_m T)\lambda^s.$$

Given $\varepsilon > 0$ **, let** $q_\varepsilon(\lambda)$ **such that for** $q > q_\varepsilon(\lambda)$,

$$\sum_{s=q+1}^{\infty} \frac{1}{s!} \left[\mathcal{J}_1^n(T)\right]^s |\lambda|^s < \varepsilon.$$

We assume that $q_\varepsilon(\lambda)$ **is large enough that for** $m > q_\varepsilon(\lambda)$

$$|\tilde{\Delta}(\lambda) - \tilde{\Delta}^{(m)}(\lambda)| < \varepsilon.$$

One has:

$$|\tilde{\Delta}(\lambda) - \Delta_m(\lambda)| \leq |\tilde{\Delta}(\lambda) - \tilde{\Delta}^{(m)}(\lambda)| + \left|\sum_{s=m+1}^{\infty} (-1)^s \mathcal{J}_s^n(P_m T)\lambda^s\right|.$$

Thus, for $m > q_\varepsilon(\lambda)$,

$$|\tilde{\Delta}(\lambda) - \Delta_m(\lambda)| < 2\varepsilon.$$

It follows that:

(18.7) $$\Delta(\lambda) = \sum_{s=0}^{\infty} (-1)^s \mathcal{J}_s^n(T)\lambda^s.$$

The function $\Delta(\lambda)$ vanishes if and only if $\lambda = \mu_k^{-n}$; μ_k^{-n}, as a zero of $\Delta(\lambda)$, has the same multiplicity that μ_k has as an eigenvalue of T.

From (18.7) it follows that the same entire function $\Delta(\lambda)$ corresponds to both the operators T and R. Then T and R have the same eigenvalues with the same multiplicities. Let:

$$Ru = \sum_{h=1}^{\infty} \mu_h (u, v_h) v_h$$

be the spectral representation of R. The operator:

$$Uv = \sum_{h=1}^{\infty} (v, u_h) v_h$$

is a unitary operator of the space S such that equation (18.6) is satisfied.

<u>Remark</u>. The function $\Delta(\lambda)$ is well known in the theory of integral equations as the "Fredholm entire trascendental function".

18.V. <u>If</u> $T_1 < T_2$ <u>then</u> $\mathfrak{J}_s^n(T_1) \le \mathfrak{J}_s^n(T_2)$ <u>for any</u> s <u>and any</u> n.

This lemma is an immediate consequence of (18.3) and of lemma (15.XI).

18.VI. <u>If</u> $T_k \rightrightarrows T$ <u>then for any</u> s <u>and any</u> n

$$\lim_{k \to \infty} \mathfrak{J}_s^n(T_k) = \mathfrak{J}_s^n(T).$$

The proof follows readily from (18.3) and from lemma 15.IX.

Bibliography of Lecture 18

[1] G. FICHERA - Funzioni analitiche di una variabile complessa - Ediz. Veschi - Roma, 1959.

[2] E. GOURSAT - Cours d'Analyse Mathématique - vol.III - Gauthier-Villars - Paris, 1924.

[3] H. HAHN - Ueber die Integrale des Herrn Hellinger und die Orthogonalinvarianten der quadratischen Formen von unendlich vielen Varändlichen - Monat.Math. Phy. Bd. 23, 1912.

[4] E. HELLINGER - Die Orthogonalinvarianten quadratischen Formen von undendlich vielen Variablen - Dissertation Göttingen, 1907.

Lecture 19

Upper approximation of the eigenvalues of a PCO. Representation of orthogonal invariants.

Let us consider an arbitrary complete system $\{w_k\}$ of linearly independent vectors in the space S. Let W_ν be the ν-dimensional manifold spanned by $w_1, \dots, w_\nu$ and P_ν the projector on W_ν. As we saw in lecture 15, the approximating eigenvalues given by the Rayleigh-Ritz method coincide with the positive eigenvalues of the operator $P_\nu T P_\nu$. If T – as we have assumed – is strictly positive, then the determinantal equation:

$$\det\{(Tw_i, w_j) - \mu(w_i, w_j)\} = 0 \qquad (i, j = 1, \dots, \nu)$$

has ν positive roots $\mu_1^{(\nu)} \geq \mu_2^{(\nu)} \geq \dots \geq \mu_\nu^{(\nu)}$.

Let us consider the eigenvectors of $P_\nu T P_\nu$ corresponding to the above eigenvalues, say $w_1^{(\nu)}, \dots, w_\nu^{(\nu)}$. Denote by $W_\nu^{(k)}$ the $(\nu-1)$-dimensional subspace of W_ν spanned by $w_1^{(\nu)}, \dots, w_{k-1}^{(\nu)}, w_{k+1}^{(\nu)}, \dots, w_\nu^{(\nu)}$. Let $P_\nu^{(k)}$ be the projector of S onto $W_\nu^{(k)}$.

The following theorem holds:

19.I. <u>Suppose</u> $T \in \mathcal{T}^m$. <u>Fixed</u> $n > 0$ <u>and</u> $s > 0$, <u>set for</u> $\nu \geq s$:

$$(19.1) \qquad \sigma_\kappa^{(\nu)} = \left\{ \frac{\mathcal{J}_s^n(T) - \mathcal{J}_s^n(P_\nu T P_\nu)}{\mathcal{J}_{s-1}^n(P_\nu^{(\kappa)} T P_\nu^{(\kappa)})} + [\mu_\kappa^{(\nu)}]^n \right\}^{\frac{1}{n}} .$$ [1]

<u>We have:</u>

$$(19.2) \qquad \sigma_\kappa^{(\nu)} \geq \sigma_\kappa^{(\nu+1)} \qquad (19.3) \qquad \lim_{\nu \to \infty} \sigma_\kappa^{(\nu)} = \mu_\kappa ,$$

<u>where</u> $\{\mu_\kappa\}$ - <u>as usual - denotes the sequence of the eigenvalues of</u> T.

We have for (18.3):

$$(19.4) \qquad \mathcal{J}_s^n(P_\nu T P_\nu) = \sum_{h_1 < \dots < h_s}^{1,\dots,\nu} [\mu_{h_1}^{(\nu)} \cdots \mu_{h_s}^{(\nu)}]^n$$

$$(19.5) \qquad \mathcal{J}_{s-1}^n(P_\nu^{(\kappa)} T P_\nu^{(\kappa)}) = \sum_{h_1 < \dots < h_{s-1}}^{1,\dots,\nu} {}^{(\kappa)} [\mu_{h_1}^{(\nu)} \cdots \mu_{h_{s-1}}^{(\nu)}]^n ,$$

where $\sum\limits_{h_1 < \dots < h_{s-1}}^{1,\dots,\nu} {}^{(\kappa)}$ means that the summation is extended to any set of $s-1$ increasing integers chosen amongst $1, \dots, \kappa-1, \kappa+1, \dots, \nu$.

It follows that:

$$[\sigma_\kappa^{(\nu)}]^n = \mu_\kappa^n \frac{\sum\limits_{h_1 < \dots < h_{s-1}}^{(\kappa)} [\mu_{h_1} \cdots \mu_{h_{s-1}}]^n}{\sum\limits_{h_1 < \dots < h_{s-1}}^{1,\dots,\nu} {}^{(\kappa)} [\mu_{h_1}^{(\nu)} \cdots \mu_{h_{s-1}}^{(\nu)}]^n} +$$

---

(1) $\sigma_\kappa^{(\nu)}$ also depends on s and on n, but we do not need to put into evidence this dependence since we consider s and n to be fixed.
For the definition of the orthogonal invariants of $P_\nu T P_\nu$ and $P_\nu^{(\kappa)} T P_\nu^{(\kappa)}$ see footnote (1) of lecture 18.

$$+ \quad \frac{\sum\limits_{h_1<\cdots<h_s}^{(K)} [\mu_{h_1}\cdots\mu_{h_s}]^n - \sum\limits_{h_1<\cdots<h_s}^{(K)\,1,\dots,\nu} [\mu_{h_1}^{(\nu)}\cdots\mu_{h_s}^{(\nu)}]^n}{\sum\limits_{h_1<\cdots<h_{s-1}}^{(K)\,1,\dots,\nu} [\mu_{h_1}^{(\nu)}\cdots\mu_{h_{s-1}}^{(\nu)}]^n}.$$

Since (see lemma 15.XII):

$$\sum_{h_1<\cdots<h_{s-1}}^{(K)\,1,\dots,\nu} [\mu_{h_1}^{(\nu)}\cdots\mu_{h_{s-1}}^{(\nu)}]^n \leq \sum_{h_1<\cdots<h_{s-1}}^{(K)\,1,\dots,\nu+1} [\mu_{h_1}^{(\nu+1)}\cdots\mu_{h_{s-1}}^{(\nu+1)}]^n,$$

(19.6)

$$\sum_{h_1<\cdots<h_s}^{(K)\,1,\dots,\nu} [\mu_{h_1}^{(\nu)}\cdots\mu_{h_s}^{(\nu)}]^n \leq \sum_{h_1<\cdots<h_s}^{(K)\,1,\dots,\nu+1} [\mu_{h_1}^{(\nu+1)}\cdots\mu_{h_s}^{(\nu+1)}]^n,$$

(19.2) follows.

From inequality (19.6) we have for $\nu \geq s$

$$\mathfrak{J}_{s-1}^{n} (P_\nu^{(K)} T P_\nu^{(K)}) \geq \mathfrak{J}_{s-1}^{n} (P_s^{(K)} T P_s^{(K)}).$$

On the other hand, one has (see lemma 15.X and 18. VI):

$$\lim_{\nu\to\infty} \mathfrak{J}_s^n (P_\nu T P_\nu) = \mathfrak{J}_s^n (T).$$

Hence, by theorem 15.XIII, (19.3) follows.

Theorem 19.I provides a sequence $\{\sigma_K^{(\nu)}\}$ which converges by decreasing to μ_K, <u>provided some orthogonal invariant</u> $\mathfrak{J}_s^n(T)$ <u>of</u> T

is known. In fact, on the right hand side of (19.1) everything but $\mathfrak{I}_s^n(T)$ is expressible in terms of the Rayleigh-Ritz approximating eigenvalues (see (19.4) and (19.5)).

Theoretically, the invariant could be considered as known since (18.2) gives the value of $\mathfrak{I}_s^n(T)$ as the sum of a series whose terms can be computed. However, it must be pointed out that partial sums of this series give a lower bound to $\mathfrak{I}_s^n(T)$, while formula (19.1) requires an upper bound for such an orthogonal invariant.

In conclusion, (18.2) can be used only when one knows how to estimate the remainder of the series on the right hand side of this formula.

In any case, theorem (19.I) can be considered as a remarkable advance in the problem of the upper approximation of the eigenvalues of T, since the upper approximation of a sequence of numbers - the μ_K's - is reduced to the upper approximation of a single number, i. e. $\mathfrak{I}_s^n(T)$ (with s and n chosen arbitrarily).

In numerical applications, the upper bound $\sigma_K^{(\nu)}$ for μ_K can be improved if we know an operator R such that $\mu_K \geq \eta_K$, for any K, where $\{\eta_K\}$ is the sequence of eigenvalues of R. Suppose we know the numerical values of the η_h's for $h=1,\dots,p$ $(p<+\infty)$. We denote by $\{z_h\}$ the corresponding eigenvectors. Let $K \leq \nu < p$ and consider the following operators:

$$R_{\nu p} u = \sum_{h=1}^{\nu} \max\{\eta_h, \mu_h^{(\nu)}\}(u, z_h) z_h + \sum_{j=\nu+1}^{p} \eta_j (u, z_j) z_j \quad .$$

$$R_{\nu p}^{(K)} u = R_{\nu p} u - \max(\eta_K, \mu_K^{(\nu)})(u, z_K) z_K \, .$$

Set :

$$\tilde{\sigma}_{k}^{(\nu)} = \left\{ \frac{\mathcal{J}_s^n(T) - \mathcal{J}_s^n(R_{\nu p})}{\mathcal{J}^n_{s-1}(R^{(k)}_{\nu p})} + [\mu_k^{(\nu)}]^n \right\}^{\frac{1}{n}}.$$

The value $\tilde{\sigma}_k^{(\nu)}$ furnishes a better upper bound to μ_k than $\sigma_k^{(\nu)}$.

We wish now, by representing the space S as a Hilbert function space, to establish some formulas which provide the calculation of $\mathcal{J}_s^n(T)$ if T is expressed as an integral operator.

Let A be a measurable set of a given measure space where a real non-negative measure μ has been introduced. Suppose that S is the Hilbert space $\mathcal{L}^2(A,\mu)$ of complex valued functions on the set A. Since S is separable, we must suppose that the measure μ is also , i.e. the σ-ring of the measurable sets is generated by a countable, semi-ring of subsets of the space. Of course the assumption that the space $\mathcal{L}^2(A,\mu)$ is the space S is not restrictive, since any abstract separable infinite dimensional Hilbert space is Hilbert-isomorphic to some space $\mathcal{L}^2(A,\mu)$. The scalar product is now given by:

$$(u,v) = \int_A u\bar{v}\, d\mu .$$

Let us suppose that T is a PCO of the class $\mathcal{C}^n$. If $\{\mu_k\}$ is the sequence of eigenvalues of T , then the operator $T^{\frac{n}{2}}$ admits the following integral representation:

$$T^{\frac{n}{2}} u = \int_A K^{(\frac{n}{2})}(x,y)\, u(y)\, d\mu_y$$

where the kernel $K^{(\frac{n}{2})}(x,y)$ is the function of $\mathcal{L}^2[A\times A, \mu\times\mu]$ defined by the development:

$$K^{(\frac{n}{2})}(x,y) = \sum_{K=1}^{\infty} \mu_K^{\frac{n}{2}} u_K(x)\,\overline{u_K(y)}\,.$$

Here, $\{u_K(x)\}$ is an orthonormal complete sequence of eigenfunctions of the operator T. Then $T^n u$ has the integral representation:

$$T^n u = \int_A K^{(n)}(x,y)\,u(y)\,d\mu_y$$

where :

$$K^{(n)}(x,y) = \sum_{K=1}^{\infty} \mu_K^{n} u_K(x)\,\overline{u_K(y)}. \tag{19.7}$$

Let us consider the function:

$$f(x_1,\dots,x_s) = \begin{vmatrix} K^{(n)}(x_1,x_1) & \dots & K^{(n)}(x_1,x_s) \\ \vdots & & \vdots \\ K^{(n)}(x_s,x_1) & \dots & K^{(n)}(x_s,x_s) \end{vmatrix}.$$

It is summable on the set $A_1\times A_2\times\dots\times A_s$. In fact we have:

$$f(x_1,\dots,x_s) = \sum_{q_1\dots q_s} \delta^{q_1\dots q_s}_{1\dots s} K^{(n)}(x_1,x_{q_1})\dots K^{(n)}(x_s,x_{q_s}).$$

Then:

$$|f(x_1,\dots,x_s)| \le s! \int_A |K^{(\frac{n}{2})}(x_1,y)|^2 d\mu_y \dots \int_A |K^{(\frac{n}{2})}(x_s,y)|^2 d\mu_y .$$

19.II. __The orthogonal invariant__ $\mathfrak{I}^n_s(T)$ __has the following integral representation:__

(19.8) $$\mathfrak{I}^n_s(T) = \frac{1}{s!} \int_A \dots \int_A f(x_1,\dots,x_s)\, d\mu_{x_1} \dots d\mu_{x_s} .$$

By using (19.7) one has:

$$f(x_1,..,x_s) = \lim_{q\to\infty} \sum_{i_1,\dots,i_s}^{1,\dots,q} \mu^n_{i_1}\dots\mu^n_{i_s} \begin{vmatrix} u_{i_1}(x_1)\overline{u_{i_1}(x_1)} & \dots & u_{i_s}(x_1)\overline{u_{i_s}(x_s)} \\ \cdot & \cdot & \cdot \\ \cdot & \cdot & \cdot \\ u_{i_s}(x_s)\overline{u_{i_1}(x_1)} & \dots & u_{i_s}(x_s)\overline{u_{i_s}(x_s)} \end{vmatrix} =$$

$$= \lim_{q\to\infty} \sum_{i_1,\dots,i_s}^{1,\dots,q} \mu^n_{i_1}\dots\mu^n_{i_s} u_{i_1}(x_1)\dots u_{i_s}(x_s) \sum_{K_1\dots K_s} \delta^{K_1\dots K_s}_{1\dots s} \overline{u_{i_{K_1}}(x_1)}\dots\overline{u_{i_{K_s}}(x_s)}.$$

Integrating over $\underset{1}{A} \times \underset{2}{A} \times \cdots \times \underset{s}{A}$ and interchanging the integral with the $\lim_{q \to \infty}$ (this is obviously possible), we get (19.8).

Remark. Representation (19.8) could be obtained by using the previously considered development of the Fredholm function $\Delta(\lambda)$ in a power series, and employing the results of the classical Fredholm theory of integral equations.

It is worthwhile, in view of the applications which we shall consider, to write down explicity formula (19.8) in the case where the measure μ is the product of the ordinary Lebesgue measure in X^n and of an atomic measure composed of m unit masses on the integers $1, 2, \cdots, p$. Then by denoting by $u_i(x)$ the value of $u(x,i)$ for $x \in A \subset X^n$ and $i = 1,2,..,p$ and, accordingly, by $K_{ij}^{(m)}(x,y)$ the kernel $K^{(m)}(x,i;y,j)$, formula (19.8) becomes:

$$(19.9) \quad \mathcal{J}_s^m(T) = \frac{1}{s!} \sum_{i_1,\dots,i_s}^{1,\dots,p} \int_A \cdots \int_A \begin{vmatrix} K_{i_1 i_1}^{(m)}(x_1,x_1) \dots \dots K_{i_1 i_s}^{(m)}(x_1,x_s) \\ \dots\dots\dots\dots\dots \\ \dots\dots\dots\dots\dots \\ K_{i_s i_1}^{(m)}(x_s,x_1) \dots K_{i_s i_s}^{(m)}(x_s,x_s) \end{vmatrix} dx_1 \dots dx_s .$$

Assuming in (19.1) that $m = 2$, $s = 1$ and inserting into it (19.9), in the particular case $p = 1$, one gets a formula already known to Trefftz [4]. He, however, even in this particular case, did not prove the convergence of $\{\sigma_K^{(\nu)}\}$ to μ_K.

Theorems 19.I and 19.II solve completely the problem of the upper approximation of a PCO T belonging to some $\mathcal{C}^n$, when an integral representation for T is known. This is the case for the eigenvalue problems arising in the classical theory of integral equations or for boundary value problems in which the corresponding Green's function or matrix is known. Unfortunately, even if we know that this Green's transformation exists - as in the case of the elliptic boundary value problems,- we are not able to compute the corresponding orthogonal invariants by using theorem 19.II, since the "kernel" of the transformation is not, in general, known.

The following theorem will be useful in attempting to overcome this difficulty.

19.III. <u>Let</u> $\{T_\rho\}$ <u>be a decreasing sequence of PCO's converging uniformly to</u> T . <u>Set:</u>

$$\sigma_k^{(\rho,\nu)} = \left\{ \frac{\mathcal{J}_s^n(T_\rho) - \mathcal{J}_s^n(P_\nu T P_\nu)}{\mathcal{J}_{s-1}^n(P_\nu^{(k)} T P_\nu^{(k)})} + [\mu_k^{(\nu)}]^n \right\}^{\frac{1}{n}}. \tag{19.10}$$

<u>We have:</u>

$$\sigma_k^{(\rho,\nu)} \geq \sigma_k^{(\tilde\rho,\tilde\nu)} \quad \text{if} \quad \rho \geq \tilde\rho \;, \quad \nu \geq \tilde\nu \tag{19.11}$$

<u>and:</u>

$$\lim_{\substack{\rho\to\infty \\ \nu\to\infty}} \sigma_k^{(\rho,\nu)} = \mu_k . \tag{19.12}$$

The proof of (19.11) follows easily if we observe that $\sigma_k^{(\rho,\nu)} \geq \sigma_k^{(\tilde\rho,\nu)}$

(lemma 18.V) and that, by the same arguments used in the proof of (19.2), $\sigma_\kappa^{(\rho,\nu)} \geq \tilde{\sigma}_\kappa^{(\tilde{\rho},\tilde{\nu})}$. (19.12) follows from $\lim_{\rho\to\infty} \sigma_\kappa^{(\rho,\nu)} = \sigma_\kappa^{(\nu)}$ (lemma 18.VI) and from (19.3).

Remark. If we replace in the formula (19.10) the operator T by T_ρ and $\mu_\kappa^{(\nu)}$ by $\mu_\kappa^{(\nu,\rho)}$ (i.e. the Rayleigh-Ritz approximation, in the space W_ν, of the eigenvalue $\mu_\kappa^{(\rho)}$ of T_ρ) we still obtain an upper bound $\tilde{\sigma}_\kappa^{(\rho,\nu)}$ for μ_κ, which is better than the one given by $\sigma_\kappa^{(\rho,\nu)}$. However, in order to compute $\tilde{\sigma}_\kappa^{(\rho,\nu)}$ we must, for any ρ, compute the Rayleigh-Rotz approximations of the eigenvalues of T_ρ.

If a "base operator" T_0 is known, then as operator T_ρ we may use the ones constructed in lecture 17. However it must be remarked that, now we do not need to know the eigenvalues and the eigenvectors for T_0, but the orthogonal invariant $\mathfrak{I}_1^m(T_\rho)$, which enters in the formula (19.10).

We wish to remark that orthogonal invariants can be used in several other topics connected with eigenvalue problems, for instance in the still partially unsolved problem consisting in the computation of the multiplicity of each eigenvalue μ_κ of T. In fact, by using orthogonal invariants, it is possible to construct sequences converging to the multiplicity of each μ_κ. It is however not yet known how to give upper and lower bounds to the multiplicity, with an error less than 1. This would determine the multiplicity completely.

For an interesting application of orthogonal invariants to this problem see [1].

Another application which can be made, concerns the mini-max principle (see 15.VIII).

We leave to the reader the proof of the following theorem (where the same notation as in theorem 15.VIII is used):

19.IV. Let T belong to $\mathcal{T}^n$. A necessary and sufficient condition for the equality sign to hold in the following relation:

$$M(v_1, \ldots, v_{k-1}) \geq \mu_k$$

is that:

$$\lim_{n \to \infty} \left[\mathcal{J}_1^{2n}(R) \right]^{\frac{1}{2n}} = \mu_k,$$

where the operator R is the following:

$$Ru = Tu - \sum_{h=1}^{k-1} (Tu, v_h) v_h - \sum_{h=1}^{k-1} (u, v_h) T v_h +$$

$$+ \sum_{i,h}^{1,\ldots,k-1} (Tp_h, p_i)(u, p_h) p_i .$$

Another approach to the two-sided approximation of the eigenvalues of a PCO in integral form is due to L. De Vito [2]. His method is highly interesting from a theoretical point of view and does not require the use of the Rayleigh-Ritz approximations (2). However the iterative

---

(2) Unfortunately the mathematical interest of De Vito's results has escaped researchers working in this area, probably because of a quite incompetent review of De Vito's paper published in Mathematical Reviews.

technique needed for the application of his procedure turns out to be rather impractical from the numerical point of view.

Bibliography of Lecture 19

[1] M.P. COLAUTTI - Sul calcolo dei numeri di Betti di una varietà differenziabile, nota per mezzo di un suo atlante - Rend. di Matem. - Roma, 1963.

[2] L. DE VITO - Sul calcolo approssimato degli autovalori delle trasformazioni compatte e delle relative molteplicità - Nota I & II - Rend. Accad. Naz. Lincei, 1961.

[3] G. FICHERA - see [1] of lecture 1.

[4] E. TREFFTZ - Ueber Fehlershätzung bei Berechnung von Eigenwerten - Math. Annalen Bd. 108, 1933.

Lecture 20

Explicit construction of the Green's matrix for an elliptic system.

Let us consider the $n \times n$ matrix differential operator of order $2m$

$$L(x,D) \equiv D^p a_{pq}(x) D^q \qquad (0 \le |p| \le m, \ 0 \le |q| \le m).$$

Suppose the coefficients $a_{pq}(x)$ to be complex $n \times n$ matrices defined in the entire X^n cartesian space and belonging to C^∞.

We make the following hypotheses;

i) The operator $L(x,D)$ is elliptic for every $x \in X^n$, i.e.

$$\det a_{pq}(x) \xi^p \xi^q \neq 0 \qquad (|p| = |q| = m) \qquad (\xi \text{ real} \neq 0);$$

ii) $L(x,D)$ is formally self-adjoint, i.e.,

$$a_{pq}(x) = (-1)^{|p|+|q|} \bar{a}_{qp}(x) ;$$

iii) Consider the bilinear form:

$$B(u,v) = (-1)^p \int_A a_{pq} D^q u \, D^p v \, dx,$$

connected with the operator $L(x,D)$ in the properly regular domain A.[1]

---

[1] for the definition of properly regular domain see lecture 3.

The corresponding quadratic form $B(u,u)$ is such that:

$$(-1)^m B(u,u) \geq c \sum_{|p|=m} \int_A |D^p u|^2 dx$$

for any $u \in C^\infty$, where c is a positive constant independent of u.

A further hypothesis will require that:

iiii) <u>A fundamental matrix in the large</u> for the operator $L(x,D)$ exists.

This matrix - say $F(x,y)$ - is defined as follows: $F(x,y)$ is a $n \times n$ matrix defined for $(x,y) \in (X^n \times X^n) - \delta$, where δ is the diagonal of the cartesian product $X^n \times X^n$ and is such that:

1) $F(x,y)$ is C^∞ in the set $(X^n \times X^n) - \delta$;

2) $F(x,y) = \overline{F(y,x)}$;

3) $D^\alpha_x F(x,y) = O\left(|x-y|^{2m-n-|\alpha|} \log |x-y|\right)$;

4) For any f belonging to $\mathcal{L}^2(X^n)$ and vanishing outside of a bounded set, the function:

$$u(x) = \int_{X^n} f(y) F(x,y) dy \tag{20.1}$$

is an $\mathcal{L}^2$-weak solution of the differential equation $Lu = f$.

From the theory of elliptic linear differential operators,[2] it follows that the function $u(x)$ has $\mathcal{L}^2$ strong partial derivatives up

---

(2) See lecture 5.

to the order $2m$ in any bounded domain of the plane.

Derivatives of order not exceeding $2m-1$ can be computed by differentiating (20.1) under the integral sign. This follows from hypothesis 3).

If $f \in C^\infty$, then $u(x)$ is C^∞ and is a solution of the differential equation $Lu = f$ in the classical sense.

In the case of elliptic operators with constant coefficients, the matrix $F(x,y)$ can be given in closed form. For instance, let us suppose that $a_{pq}(x) \equiv 0$ for $|p|+|q|<2m$ and denote (for $|p|=|q|=m$) by a_{pq} constant matrices such that $Q(\xi) = \det a_{pq}\xi^p\xi^q \neq 0$. Let $L(D) = a_{pq}D^pD^q u$. By $\tilde{L}(\xi)$ we shall denote the matrix obtained by taking the matrix of the cofactors of $a_{pq}\xi^p\xi^q$ and transposing it. Let us denote by Ω_ξ the unit sphere $|\xi| = 1$ in the r-dimensional cartesian space and by $d\omega_\xi$ the measure of the hyper-surface element on Ω_ξ. Define (Δ_2 is the Laplace operator)

$$S(x,y) = \frac{1}{4(2\pi i)^{r-1}(2m-1)!}(\Delta_{2y})^{\frac{r-1}{2}} \int_{\Omega_\xi} \frac{|(x-y)\cdot\xi|^{2m-1}}{Q(\xi)}\, d\omega_\xi$$

for r odd, and

$$S(x,y) = \frac{-1}{(2\pi i)^r(2m)!}(\Delta_{2y})^{\frac{r}{2}} \int_{\Omega_\xi} \frac{|(x-y)\xi|^{2m}\log|(x-y)\xi|}{Q(\xi)}\, d\omega_\xi$$

for n even (see [6]). Then $F(x,y)$ is defined as follows:

$$F(x,y) = \tilde{L}(D)\,S(x,y).^{(3)}$$

In the general case of variable coefficients the existence of a fundamental solution in the large has been proven by Giraud [4] for $n=1$, $m=1$. For m arbitrary, see [3]. The method described in that paper can be extended to the case of systems.

Let us consider in the space $H_m(A)$ (space of vector valued functions with $\mathcal{L}^2$ strong derivatives up to the order m) the new scalar product:

$$((u,v)) = (-1)^m B(u,v).$$

The space obtained from $H_m(A)$ by functional completion with respect to this new scalar product will be denoted by $\mathcal{H}(A)$. If we denote by Γ the finite dimensional vector space composed of polynomials w of degree $\leq m-1$, such that $B(w,w)=0$, we must consider two functions of $\mathcal{H}(A)$ as coinciding when they differ by a polynomial belonging to Γ. The space $\mathcal{H}(A)$ is none other, except for a Hilbert isomorphism, than the quotient space $H_m(A)/\Gamma$.

Let us denote by $(\ ,\ )$ the scalar product in $\mathcal{L}^2(A)$ and consider the operator

$$Ru = \int_A u(y)\,F(x,y)\,dy.$$

---

(3) If $\tilde{L}_{ij}(\xi)$ is the element of $\tilde{L}(\xi)$, then by $\tilde{L}(D)S$ we mean the matrix whose elements are $\tilde{L}_{ij}(D)S$.

Since for $u \in \mathcal{L}^2(A)$, Ru belongs to $H_m(A)$, we can consider R as an operator with domain $\mathcal{L}^2(A)$ and range in the space $\mathcal{H}(A)$. It is easily seen that R is a compact operator. The adjoint operator R^* has $\mathcal{H}(A)$ as a domain and range in $\mathcal{L}^2(A)$. For $v \in H_m(A)$ it is expressed as follows:

$$R^* v = (-1)^{|q|+m} \int_A (a_{qp}(x) D_x^p v(x)) D_x^q F(y,x)\, dx$$

and represents a function belonging to H_m in any compact set of the plane.

Let us now consider the following BVP:

(20.2) $L(x,D)u = (-1)^m f$ in A, (20.3) $D^p u = 0$ on ∂A

$0 \leq |p| \leq m-1$.

Suppose we wish to represent the solution in the following way: $u = R^* v$ by a proper choice of v. Let A_0 be a bounded domain containing $\bar{A}$ in its interior. Let $\{\varphi_k(x)\}$ be a complete system in the space $\mathcal{L}^2(A_0 - A)$. The boundary conditions (20.3) will be satisfied (in the sense of functions of $H_m(A)$) if the function $R^* v$ considered for $y \in A_0 - A$ is such that:

(20.4)
$$\int_{A_0 - A} \varphi_k(y) R^* v \, dy = 0 \qquad (k = 1,2,\dots)$$

Set:

$$\omega_k(x) = \int_{A_0 - A} \varphi_k(y) F(x,y)\, dy.$$

Conditions (20.4) can be written:

$$B(\omega_k, v) = 0 \qquad (k = 1, 2, \dots). \tag{20.5}$$

Let us consider the manifold V of solutions of the homogeneous equation $L(x, D)u = 0$ in A, which belong to $\mathcal{H}(A)$. This manifold is a closed subspace of $\mathcal{H}(A)$, since it is a closed subspace of $H_m(A)$.

Let P be the projector of $\mathcal{H}(A)$ onto V. Condition (20.5) is satisfied if $Pv = 0$. It follows that for any $v \in H_m(A)$, the function $u = R^*(v - Pv)$ satisfies the boundary conditions (20.3). For $v \in H_m(A) \cap H_{2m}(A')$ (for every A' such that $\bar{A}' \subset A$) we have:

$$L R^* v = (-1)^m L v$$

as can be proved easily. It follows that the function:

$$u = R^* R f - R^* P R f$$

is the solution of the BVP (20.2), (20.3).

We have thus constructed explicitly the Green's transformation:

$$\boxed{G = R^* R - R^* P R.}$$

This construction is perfectly suitable for applying theorem 19.III by using results of lecture 17.

In fact, let us take a basis in the subspace V, say $\{\omega_k\}$, and denote by P_ϱ the projector of $\mathcal{H}(A)$ onto the subspace spanned by $\omega_1, \ldots, \omega_\varrho$. From theorem 17.III it follows that $G_\varrho = R^* R - R^* P_\varrho R$ converges uniformly to G. On the other hand, the operator G and the operators G_ϱ belong to $\mathcal{C}^n$ for any n such that:

$$n > \frac{r}{2m} .$$

This follows from property 3) of $F(x, y)$.

The orthogonal invariants of G_ϱ corresponding to such n can be calculated by using the results of lecture 19, since $R^* R$ can be expressed as an integral operator and the same is true for $R^* P_\varrho R$, which is a degenerate operator.

It follows that we may consider as solved the eigenvalue problems connected with the boundary value problem (20.2), (20.3), i.e.

$$G\varphi - \mu\varphi = 0.$$

Let us consider some particular cases corresponding to classical eigenvalue problems of mathematical Physics. For these problems we shall construct explicitly the approximating sequences for the eigenvalues.

Let us first consider the classical operator of linear elasticity for an isotropic homogeneous body, which we write as follows in the space X^2 or X^3:

$$L_i u \equiv u_{i/hh} + \alpha u_{h/ih} = f_i \qquad (\alpha > -1) \quad \text{in } A ,$$

with the boundary condition $u = 0$ on ∂A. As bilinear form we may assume the following:

$$B(u,v) = -\int_A (u_{i/h} v_{i/h} + \alpha u_{i/i} v_{h/h})\, dx,$$

(we consider from now on only real vector-valued functions).

Let us assume that:

$$\varphi(t) \begin{cases} = \log t^{-1} & \text{for } r = 2, \\ = t^{-1} & \text{for } r = 3. \end{cases}$$

The fundamental matrix - as given by Somigliana - is the following:

$$F_{ij}(x-y) = \frac{\alpha}{8\pi(1+\alpha)} \frac{\partial^2 |x-y|^2 \varphi(|x-y|)}{\partial x_i \partial x_j} - \frac{\delta_{ij}}{(r-1)2\pi} \varphi(|x-y|).$$

Set:

$$\gamma_{ij}(x,y) = -\int_A \left\{ F_{ik/h}(x-t) F_{jk/h}(t-y) + \alpha F_{ik/k}(x-t) F_{jh/h}(t-y) \right\} dt.$$

Consider a complete system $\{\omega^p\}$ of solutions of the homogeneous equation $Lu = 0$, such that $-B(\omega^p, \omega^q) = \delta_{pq}$.[4] Set:

---

[4] For the construction of complete systems of solutions of the equation $Lu = 0$ see [4] chap. III.

$$\rho_i^p(x) = \int_A \left\{ F_{ik/h}(x-t)\,\omega^p_{k/h}(t) + \alpha\, F_{ik/k}(x-t)\,\omega^p_{h/h}(t) \right\} dt .$$

For the eigenvalues $\{\lambda_k\}$ of the following problem:

$$Lu + \lambda u = 0 \quad \text{in } A, \qquad u = 0 \quad \text{on } \partial A ,$$

we have:

$$\frac{1}{\mu_k^{(\nu)}} \geq \lambda_k \geq \tau_k^{(\nu)}$$

$$\lim_{\nu \to \infty} \frac{1}{\mu_k^{(\nu)}} = \lim_{\nu \to \infty} \tau_k^{(\nu)} = \lambda_k .$$

The $\mu_k^{(\nu)}$ are the roots of the following determinantal equation:

$$\det \left\{ (w_i, w_j) + \mu (w_i, L w_j) \right\} = 0 \qquad (i,j = 1,\dots,\nu)$$

where $\{w_i\}$ is any complete system of functions vanishing on ∂A.
The $\tau_k^{(\nu)}$ are given by the following formula:

$$\tau_k^{(\nu)} = \left\{ \sum_{ij}^{1,r} \int_A \int_A |\gamma_{ij}(x,y)|^2 dx\,dy + \sum_{pq}^{1,\nu} \int_A \rho_i^p(x)\,\rho_i^q(x)\,dx \int_A \rho_j^p(x)\,\rho_j^q(x)\,dx - \right.$$

$$\left. -2 \sum_p^{1,\nu} \int_A \int_A \gamma_{ij}(x,y)\,\rho_i^p(x)\,\rho_j^p(y)\,dx\,dy - \sum_{i=1}^{\nu}{}^{(k)} \left[\mu_i^{(\nu)}\right]^2 \right\}^{-\frac{1}{2}}$$

It is easy for $r = 2$ and $\alpha = 2$ to derive from the above formulas the approximations for the eigenvalues for a membrane fixed along its boundary.

As a second example, let us consider the vibrations of a plate clamped along its boundary, i.e. the two-dimensional eigenvalue problem

$$\Delta_2 \Delta_2 u - \lambda u = 0 \quad \text{in } A \quad , \qquad u = \frac{\partial u}{\partial n} = 0 \quad \text{on } \partial A .$$

In this case, the lower bounds $\tau_k^{(\nu)}$ are expressed, by means of the Rayleigh-Ritz approximations, as follows:

$$\tau_k^{(\nu)} = \left\{ \frac{1}{4\pi^2} \int_A \int_A | \log |x-y| |^2 \, dx \, dy - \right.$$

$$\left. - \frac{1}{4\pi^2} \sum_{h=1}^{\nu} \int_A \left[\int_A \omega_h(t) \log |x-t| \, dt \right]^2 dx - \sum_{i=1}^{\nu} {}^{(k)} \mu_i^{(\nu)} \right\}^{-1} .$$

$\{ \omega_h \}$ is an orthonormal system of harmonic polynomials in $\mathcal{L}^2(A)$; A is supposed simply connected.

As a last example, let us consider the eigenvalue problem connected with the buckling of a clamped plate:

$$\Delta_2 \Delta_2 u + \lambda \Delta_2 u = 0 \quad \text{in } A \quad , \qquad u = \frac{\partial u}{\partial n} = 0 \quad \text{on } \partial A .$$

After computing the Rayleigh-Ritz approximation, we have for the lower approximation of λ_k :

$$\tau_{\kappa}^{(\nu)} = \frac{1}{4\pi^2}\left\{\int_A\int_A |\log|x-y||^2 dx\,dy - 2\sum_{j=1}^{\nu}\int_A\left[\int_A \omega_j(t)\log|x-t|\,dt\right]^2 dx + \right.$$

$$\left. + \sum_{h\,j}^{1,\nu}\left(\int_A\int_A \log|x-y|\,\omega_h(x)\,\omega_j(y)\,dx\,dy\right)^2 - \sum_{i=1}^{\nu}{}^{(\kappa)}\left[\mu_i^{(\nu)}\right]^2\right\}^{-\frac{1}{2}}.$$

The ω_h have the same meaning as in the previous example.